Philip Schech

Diseases of the Mouth, Throat, and Nose

Philip Schech

Diseases of the Mouth, Throat, and Nose

ISBN/EAN: 9783744693943

Printed in Europe, USA, Canada, Australia, Japan

Cover: Foto ©berggeist007 / pixelio.de

More available books at **www.hansebooks.com**

OF THE

MOUTH, THROAT, AND NOSE.

INCLUDING

RHINOSCOPY AND METHODS OF LOCAL TREATMENT.

BY

Dr. PHILIP SCHECH,
LECTURER IN THE UNIVERSITY OF MUNICH.

TRANSLATED BY

R. H. BLAIKIE, M.D., F.R.C.S.E.,
SURGEON EDINBURGH EAR AND THROAT DISPENSARY, LATE CLINICAL ASSISTANT
EAR AND THROAT DEPARTMENT, ROYAL INFIRMARY, EDINBURGH.

WITH FIVE ILLUSTRATIONS.

EDINBURGH:
YOUNG J. PENTLAND.
1886.

AUTHOR'S PREFACE.

So rapidly has our knowledge of Oral and Pharyngeal Diseases, and particularly of Diseases of the Nose,—the significance of which, however, has not as yet been sufficiently appreciated,—extended of late, that it is now impossible, either for the practitioner or the student, to avoid bestowing on them a measure of attention. There are certainly many good works and writings on these subjects already, but most of them seem to be either too voluminous or too short. I have therefore very gladly complied with the request of the publishers to write a short but exhaustive treatise on the subjects, all the more because, having worked for twelve years at the literature as well as the practice of these diseases, I have now come to know exactly what the practitioner requires. The arrangement may appear unusual to some, but I think it is as good as any other. Throughout the work great stress has been laid upon the objective condition. This has appeared to me absolutely necessary, on account of the great number of diseases which have similar symptoms; in fact, in diseases of the Mouth, Pharynx, Larynx, and Nose, one is almost compelled to base the diagnosis on the objective symptoms alone. The more detailed discussion in some chapters is explained by the frequency and importance of the affection described. Whoever inclines to think the treatment described in too special a manner, should bear in mind that no simpler, and at the same time efficient, method is possible. The reader will find the services of others suitably acknowledged, but at the same time will not miss the independent experience and criticism of the author.

Dr PHILIP SCHECH.

PREFACE TO THE TRANSLATION.

This translation was undertaken with the idea of presenting to the Profession a short and concise work on the subjects it deals with. The very favourable criticisms of the book in the German Medical Journals show that it is one deserving the attention of all interested in this branch of Medical Science. It was written by Dr Schech, as a companion volume to Dr J. Gottstein's "Krankheiten des Kehlkopfes," which has been translated by Dr. M'Bride.

In the Appendix will be found a few Notes, which it is hoped will add to the value of the book. I am glad of this opportunity of thanking Dr M'Bride for the interest he has taken in the work, and his kindness in revising the translation, and Dr M'Kenzie Johnston for revising the proof sheets.

R. H. BLAIKIE.

Edinburgh, *March* 1886.

CONTENTS.

SECTION I.

DISEASES OF THE MOUTH.

PAGE

ANATOMICAL AND CLINICAL INTRODUCTORY REMARKS—METHODS OF EXAMINATION—MALFORMATIONS, ANOMALIES, AND DEFORMITIES—CATARRHAL STOMATITIS—PHLEGMONOUS AND PARENCHYMATOUS STOMATITIS—EXUDATIVE STOMATITIS—ULCERATIVE STOMATITIS—DIPHTHERITIC STOMATITIS—GANGRENOUS STOMATITIS—SYPHILIS—TUBERCULOSIS—PARASITIC DISEASES—HÆMORRHAGES—NEOPLASMS AND TUMOURS—DISEASES OF THE NERVES—DISEASES OF THE SALIVARY GLANDS—SALIVATION OR PTYALISM—PAROTITIS—NEOPLASMS AND CONCRETIONS—ANGINA LUDOVICI, OR CYNANCHE CELLULARIS MALIGNA, . . 1-68

SECTION II.

DISEASES OF THE THROAT.

ANATOMICAL AND CLINICAL INTRODUCTORY REMARKS—EXAMINATION OF THE PHARYNX—GENERAL THERAPEUTICS—MALFORMATIONS AND DEFORMITIES, CONGENITAL AND ACQUIRED, AND OTHER ANOMALIES—ACUTE CATARRHAL PHARYNGITIS—CHRONIC PHARYNGITIS—PHLEGMONOUS PHARYNGITIS—RETRO-PHARYNGEAL ABSCESS—EXUDATIVE PHARYNGITIS—DIPHTHERITIC PHARYNGITIS—GANGRENOUS PHARYNGITIS—SYPHILIS—TUBERCULOSIS—SCROFULA, LUPUS, LEPRA, AND GLANDERS—PARASITIC DISEASES—HÆMORRHAGES—FOREIGN BODIES AND CONCRETIONS—NEOPLASMS AND TUMOURS—DISEASES OF THE NERVES, . . 69-198

SECTION III.

DISEASES OF THE NOSE.

ANATOMICAL AND CLINICAL INTRODUCTORY REMARKS—METHODS OF EXAMINATION—GENERAL THERAPEUTICS—DEFORMITIES, ANOMALIES, AND MALFORMATIONS—ACUTE CATARRHAL RHINITIS—CHRONIC RHINITIS—PHLEGMONOUS RHINITIS—DIPHTHERITIC RHINITIS—SYPHILIS—TUBERCULOSIS—SCROFULA, LUPUS, AND GLANDERS — PARASITIC DISEASES — HÆMORRHAGES — FOREIGN BODIES AND CONCRETIONS—NEOPLASMS AND TUMOURS—DISEASES OF THE NERVES—DISEASES OF THE ACCESSORY CAVITIES, 199-284

APPENDIX, ... 285-292

INDEX, . . 295

LIST OF ILLUSTRATIONS.

Fig.
1. THE RHINOSCOPIC IMAGE, ... 79
2. SCHECH'S UNIVERSAL HANDLE FOR GALVANO-CAUSTIC OPERATIONS IN THE PHARYNX, NOSE, AND LARYNX, . . . 87
3. SCHECH'S POST-NASAL FORCEPS, ... 110
4. VARIOUS NASAL SPECULA, ... 210
5. WIRE SNARE FOR THE NOSE (COLD SNARE), . . . 267

SECTION I.

DISEASES OF THE MOUTH.

DISEASES OF THE MOUTH.

ANATOMICAL AND CLINICAL INTRODUCTORY REMARKS.

The mouth is that cavity bounded in front by the lips, at the sides by the mucous membrane of the cheeks, above by the hard palate, behind by the velum palatinum, and below by the tongue and those muscles that pass from the inferior maxilla to the hyoid bone. By the protrusion of the alveolar processes of the upper and under jaws, the oral cavity is divided into two parts, the smaller (vestibulum oris) in front, the larger (cavum oris proper) behind. These two parts communicate with each other by means of an interstice lying between the last molar tooth and the anterior edge of the coronoid process of the under jaw. The aperture of the oral cavity in front (rima oris) is formed by the lips, two very flexible organs, whose covering is continuous with the mucous membrane of the digestive tract, and each of which is joined to the gum by a small fold of mucous membrane (frenulum labii super. et infer.). The mucous membrane of the lips is continuous, on the one hand, with that on the inner surface of the cheeks, on the other with that of the alveolar processes of the upper and lower jaws, thence extending through the spaces between the teeth on to the hard palate, and by the lingual ligament on to the tongue and the excretory ducts of the sub-lingual and submaxillary glands. Although the anterior surface of the soft palate anatomically belongs to the oral cavity, it will be considered, for purely practical reasons, under the structures of the pharynx.

The most important constituent part of the oral cavity is the tongue. Its superior convex surface, as far back as the isthmus faucium, is supplied with organs of touch and taste, making it appear rough and ragged. To be distinguished are small thread-like elevations—the Filiform Papillæ, which are scattered over the whole superior surface in large quantities, giving it a rough furred appearance; and the Fungiform Papillæ, larger than the former, and dispersed among them like little red buttons or conically shaped knobs; and lastly, the Circumvallate Papillæ, situated in the form of a V at the isthmus faucium. The last named, rising considerably above the rest of the membrane, are often regarded by anxious people as pathological products. The most posterior part of the tongue between the circumvallate papillæ and the epiglottis has on it only a few well-developed follicular glands.

Very various appearances are met with on inspecting the superior surface of the tongue. While in most cases it is smooth and regular, it often appears furrowed and fissured, both longitudinally and transversely (rarely the latter). Frequently, under normal as well as pathological circumstances, the impression of the teeth is found on the edges of the tongue. The term cat or strawberry tongue is used when the papillæ are more prominent, as is usually the case in scarlet fever.

The mucous membrane of the dorsum of the tongue is covered with a white or whitish yellow substance, which towards the tip of the tongue gradually diminishes, or even disappears entirely, becoming thicker in the middle, however, especially towards the isthmus faucium. This is the fur or coating of the tongue, and, according to Unna, depends on the ciliated filiform papillæ. In many quite healthy persons the coating of the tongue attains a certain thickness, otherwise seen only in pathological circumstances. This pathological coating is caused in most cases, according to Unna, by the swelling of all the hair-like processes of the filiform papillæ. The white of the coating is regulated by the distention of the superior layer of the epidermis, although fungus growths also appear in it. Dryness of the upper surface of the tongue, as well as a dirty brown fur, is observed in persons suffering from fever or diabetes, or in those who cannot breathe through the nose. A black or dark brown fur may be due to medicines, food, or to coagulated blood from small fissures in the mucous membrane.

The form of the tongue coating is very varied. It appears sometimes in longitudinal striæ at the edges or in the middle of the tongue, sometimes in rounded or oval spots, sometimes in spiral lines and circles, with a red centre, and sometimes in united patches or blotches. This coating, examined through a microscope, is observed to consist principally of young or old epithelium in various stages of fatty degeneration, remnants of food, fatty molecules, mucous corpuscles, fungus fibres (*Leptothrix buccalis*), and different kinds of zoogloea, rod-shaped as well as circular.

The size of the tongue varies as much as its powers of motion. Inequality of the halves of the tongue is a sign of atrophy; inclination of the tip to one side indicates hemiplegia; and fibrillar twitching or restlessness when the tongue is protruded is a sign of paresis, anxiety, or fever.

The mucous membrane of the oral cavity is not at all places equally firmly connected with the underlying structures. It is most firmly attached over the hard palate, more loosely at the places where it passes to the gum and lips. Its epithelium is of the pavement variety, which on the tongue is constantly being shed. It is particularly rich in mucous glands, of which the excretory ducts, as small as pin points, are easily recognised, while the acini, united in groups, shine through the mucous membrane of the lips.

The tongue is specially rich in lymphatic vessels. The capillaries of the mucous membrane and of its papillæ form extensive net-works, capable of being injected, which unite to form larger branches, and anastomose with those of the pharynx and tongue. Of lymphatic glands particularly to be mentioned are the anterior auricular, the deep facial, and, above all, the submaxillary, together with the superficial and deep-lying cervical and jugular glands.

The arterial supply of the oral cavity springs from various sources. The external maxillary artery, for example, sends a submental branch for the skin and muscles under the chin, glandular and muscular branches for the parotid and submaxillary glands, the digastric, stylohyoid, masseter, and internal pterygoid muscles, and coronary branches to the upper and lower lips; the posterior auricular artery from the external carotid supplies the parotid gland, the digastric muscle, stylohyoid, styloglossus, &c.; similarly the temporal artery supplies the parotid, and by means of the transverse facial the cheek and face muscles; to the teeth go branches of the internal maxillary, the inferior dental to the teeth of the under jaw, the superior alveolar to those of the upper jaw, and, lastly, branches of the internal maxillary also supply the muscles of mastication.

The tongue, however, contains the largest vessels. The lingual artery from the external carotid arises at the level of the large horn of the hyoid bone, and, passing between the hyoglossus and middle constrictor of the pharynx, enters the tongue. Its ramifications in the tongue substance are called the hyoid, dorsalis linguæ, sublingualis and profunda linguæ or ranine; the last is the largest branch and enters the tongue near the frenulum linguæ, therefore it may be easily injured in operations on the floor of the oral cavity, as, for example, cutting the frenulum.

The veins—buccal, superior and inferior labial, masseteric, parotid, and submental—flow into the anterior facial; the transverse facial, alveolar (super. et infer.) and parotid flow into the posterior facial; and the lingual veins, arising under the tip of the tongue, into the internal jugular.

The nerves of the oral cavity contribute to the sense of touch, the sense of taste, and also to the movement of the muscles.

The nerves of sensation spring from the trigeminus, those of taste from the glossopharyngeal. A specific sense of taste is ascribed to the lingual branch of the trigeminus, particularly to the chorda tympani (which is contained in it and arises in the facial), because it extends to those regions of the mucous membrane of the tongue which the glossopharyngeal does not reach.

The mobility of the tongue is due to the hypoglossal nerve; its branches reach to the following muscles—hypoglossus, geniohyoid, genioglossus, longitudinalis and transversus linguæ. The mylohyoid muscles, as well as the anterior belly of the digastric, receive their nerve supply from the third branch of the trigeminus, the masseter from the motor portion of the trigeminus, the orbicularis oris, as also the muscles of expression, from the facial.

The muscular structure of the oral cavity consists of the muscles of the lips, cheeks, and tongue.

Among the muscles of the cheeks the most important is the

orbicularis oris. It shuts the mouth, forms the lips for whistling, kissing, and blowing, lengthens them into a long narrow channel, as in sucking.

To the muscles of expression belong the levator labii superioris, levator and depressor anguli oris, the zygomatici major and minor, the risorius Santorini, levator menti, &c.

Of the muscles of the cheek the buccinator is to be noticed. If it acts alone, it widens the oral cavity transversely; if this widening be prevented by the simultaneous acting of the orbicularis oris, it presses the cheeks on the teeth, or it contracts the oral cavity when filled with air, which, if the lips slightly part, will escape with force, as in playing on a wind instrument—hence called the trumpeter muscle.

As muscles of mastication are reckoned the temporal, the masseter, and the pterygoids (internal and external). They all have the function of drawing the lower towards the upper jaw; the external pterygoid not only raises the lower jaw, but pushes it forwards, and if one muscle only acts, towards the opposite side, being assisted by the internal pterygoid.

The floor of the oral cavity is formed by the digastric (biventer), mylohyoid, geniohyoid, stylohyoid, genioglossus, hyoglossus, and styloglossus muscles.

The tongue itself is composed of three special muscular layers, confined to it alone, as well as of the interwoven fibres of the genioglossus, hyoglossus, and styloglossus.

The superior longitudinal stratum, immediately under the mucous membrane, inserts its bundles (fasciculi) between those of the genioglossus. The inferior surpasses the superior in strength. It extends between the genioglossus and hyoglossus on the inferior surface of the tongue as far as the tip. The transverse muscular stratum —transverse lingual—arises from the lateral surfaces of the septum linguæ. Its fibres run outwards and upwards, the inner go to the dorsum of the tongue, the outer to the edge, and push themselves through between the longitudinal fibres of the genioglossus and hyoglossus.

The genioglossal muscles draw the tongue forwards and push it out, the hyoglossal draw it back. Working simultaneously, they flatten it out. Individual muscular fibres can contract independently of each other, whereby the tongue may move up and down, and to right or left, as well as form itself into an arch or a gutter.

METHODS OF EXAMINATION.

The chief method of examining the oral cavity is by inspection.

Common daylight may be allowed to fall into the oral cavity, or artificial light when the other is impracticable on account of dull weather, or at night. One should never omit to examine very carefully the teeth and gums, which is done by inserting the fore-finger into the corner of the oral cavity, and drawing back the cheeks, or turning over the lips. Great difficulty is often experienced in examining the mouth. While singers, or those acquainted with the art of singing, can, without exception, put their tongue into any desired position, inexperienced persons are most awkward, especially *bon-vivants*, stout or timid people. Such people generally, at the mere attempt to put out the tongue, have an inclination to retch or vomit, and a rotatory movement of the organ cannot be prevented. In them, also, fissures in the lingual ligament frequently result. It is well known how very difficult it is to make a laryngoscopic examination on account of the so-called rising of the tongue.

The posterior part of the tongue is best examined by the laryngeal mirror, although palpation also is of great service. The latter is indispensable in defining the extent and consistence of tumours or foreign bodies, as well as the nature of the bases and edges of ulcers.

Important conclusions are come to in many cases by means of the sense of smell, as well as by chemical analysis of secretions in the mouth, and also by microscopic examination of the fur, secretions, and constituent parts of tissues.

MALFORMATIONS, ANOMALIES, AND DEFORMITIES.

Among defects of the oral cavity are first to be considered those fissures which are formed by the individual parts of which the face is composed, being developed separately; or, in cases in which separation is normal at a certain stage of development, by their not afterwards uniting.

The best known and most common malformation is cleft or hare-lip. It usually affects the upper lip to one side of the middle line, and often extends to the ala nasi, or into the nostril. Of clefts within the mouth itself is to be mentioned fissure of the jaw between the incisor and eye-tooth. This fissure may extend to the hard palate, thus producing cleft palate, or, when occurring on both sides, "Wolfsrachen," or wolf's jaw, in which case, generally, the lips and soft palate are also divided. The most important consequence of this malformation is inability to take nourishment. As sucking is rendered difficult, the child cannot thrive; and when nourishment is given with a spoon, a portion of it is ejected through the nose. Partial or complete closure of the mouth, through union of the lips, has been observed, as also abnormal smallness of the oral opening—"Microstomia"—so that sometimes a probe can scarcely be passed. Further, shortness of the lips—"Brachylia"—double formation of the lips, abnormal fistulæ, and, lastly, abnormal connections between the lips and the alveolar edges of the jaws or palate, are sometimes seen.

Hypertrophy of the lips is relatively frequent. It is sometimes hereditary, sometimes acquired, and depends on hypertrophy of the interstitial connective tissue. The thick upper lip of scrofulous and tubercular children is well known, as also the same condition in adults, caused by repeated inflammation of these parts.

Congenital anomalies, such as the complete absence of the tongue, or its excessive or defective development, have been observed, as also connections between it and the gum or floor of the oral cavity, and deficient or excessive length of the frenum linguæ. Formerly much mischief was done while trying to remedy abnormal shortness of the frenum, and every possible defect was attributed to the supposed existence of this abnormality. It is true, however, that in cases where the frenum is inserted too near the front, the motions of the tip of the tongue are limited, and sucking and speaking rendered difficult; whereas over-length of the frenum causes the so-called swallowing of the tongue, which in sucking rises against the hard palate and produces dyspnœa.

It may be mentioned here that Schäffer once observed a nævus of the face which extended to the mucous membrane of the mouth, throat, and larynx.

As the treatment of the above anomalies, &c. belongs to the province of general surgery, it is sufficient only to mention them here.

CATARRHAL STOMATITIS.

Synonyms.—Stomatitis Catarrhalis.

Inflammation of the oral cavity is distinguished in its simplest form by erythematous reddening and swelling of the mucous membrane, the secretion of which is generally increased.

Etiologically there is a primary or idiopathic, and a secondary or symptomatic stomatitis.

Primary stomatitis may be due to mechanical, chemical, and thermal irritants, such as sharp teeth, dental caries, teething, continued difficulty in sucking in consequence of scarcity of milk in the breasts, bad nipples, too hot, cold, highly seasoned, or acid food, ingesta with sharp edges, nuts, almonds, unripe fruit, alcohol, tobacco, mineral acids, and caustic alkalies. Epstein ascribes the extraordinary predominance of diseases of the mouth in infants to the irritation of the air, the mechanical act of sucking, or the tenderness of the epithelium, and the often too energetic cleansing of the oral cavity on the part of midwives or others in charge of the child.

Secondary stomatitis accompanies most fevers and infectious diseases, such as scarlatina, measles, and small-pox, as well as most affections of the gastro-intestinal canal. And lastly, it appears after the use of certain medicines, such as iodine and mercury.

Among the subjective symptoms pain is the most prominent. The sensitiveness noticeable when the ingesta are first received increases to intense pain, especially at the tip of the tongue, when mastication is attempted. Children refuse the bottle, or, as soon as they lay hold of it, drop it, crying and screaming. Most patients complain at first of heat and dryness in the mouth, and afterwards of a superabundance of viscid secretion. Taste is considerably changed; all food, but especially that of aromatic flavour, either has no taste or a bad one—for instance coffee, which most patients reject sooner than anything else. This putrid, insipid, or clammy taste must be partly due to the decomposition of secretions, and partly to change in the gustatory nerves.

Although adults may suffer but little physically from this affection, yet it is easy to understand that infants and young children may be much weakened, and even eventually cut off, through a prolonged attack of stomatitis.

The objective symptoms consist in a more or less pronounced redness and swelling of the mucous membrane. The tongue, as also the cheeks, show impressions of the teeth; the mucous membrane is covered with a whitish, viscid, insipid, or decomposing mucus, which is very rich in cells and nuclei, and reaches a considerable thickness, especially on the tongue and gum. The papillæ of the tongue are swollen, stand out prominently, and are separated by deep furrows; their investing epithelium is much infiltrated, and of a whitish grey or yellowish colour. In rare cases, the epithelium is thrown off, and the bright red, sometimes bleeding tops of the papillæ appear (Papillitis).

If the disease be localised in the gum (Gingivitis), the discomfort is, on the whole, less serious. It (the gum) appears deep or bluish red, and the processes between the teeth are swollen and covered with a greasy secretion. Inflammation of the gum behind the upper incisor teeth and the neighbouring parts of the hard palate is very painful. The natural and abundant folds of mucous membrane found here swell to thick tubercles, which are extremely sensitive to touch, although they appear but slightly reddened. The mucous membrane of the hard palate is reticularly injected, and the mucous glands appear swollen and project visibly, sometimes as whitish grey or reddish grey papules as large as a millet seed, or as clear hemispherical vesicles.

Ulcers do not occur in catarrhal inflammation of the mouth, except in children or in bad cases of typhoid, and then at the most they only go so far as to cause shedding and loosening of the epithelium.

The course of acute idiopathic stomatitis is favourable, as the symptoms begin noticeably to disappear generally in a few days; the symptomatic form is favourable in its prognosis, its course corresponds to that of the primary disease, and is often very protracted; indeed a case of chronic stomatitis has been known to last months and even years, as long as the diseases which excited it remained.

Treatment.—The treatment must before all things be directed against the exciting causes; injurious ingesta forbidden, smoking left off, defective teeth extracted, bad breasts discontinued, and inferior bottles replaced by better ones.

The diet must be limited to fluid or quite soft substances, as milk, eggs, and soup, which, on account of the extraordinary sensitiveness of the oral cavity, should be taken lukewarm or cold. Acid or strongly saccharine fruit juices are also to be forbidden, the best fluid being pure and fresh spring water. The much recommended aërated mineral waters often cause a pricking, burning sensation, and should not therefore be much used.

The indicatio morbi renders local treatment of the diseased mucous membrane necessary. In a case of diffuse inflammation medicated mouth washes are simplest, and cause the least inconvenience to the patient. They are kept in contact with the walls of the oral cavity for the longest possible time by slightly dilating the cheeks. But one must be as careful in the choice of ingredients as in their concentration.

Chlorate of potassium in one to two[1] per cent. solutions is most to be recommended, also borax and boracic acid in one to four per cent. solutions, without the addition of honey or syrup. Weak solutions of astringents, tannin, sulphate of zinc, and alum, are also of use. In circumscribed disease, as, for example, gingivitis, with loosening of the gum, painting with glycerine of tannin (50 grs. ad ʒi.), tinct. gallar., or tinct. kram., is useful. The last two are specially useful as mouth washes, mixed with tinct. myrrh. and diluted with water (one tea-spoonful to 200 grammes).[2] Children who cannot rinse out the mouth must have the oral cavity carefully cleansed with cold water after each meal, and then washed out by a clean linen rag dipped in the solutions above mentioned, especially with borax (grs. 50 ad ʒi.) or boracic acid (grs. 10 ad ʒi.), or the applications may be made with a fine hair pencil.

---

[1] One per cent. = approximately gr. v. ad ʒi.
[2] 200 grammes = 6¼ ounces.

PHLEGMONOUS AND PARENCHYMATOUS STOMATITIS.

Synonyms—Stomatitis Phlegmonosa et Parenchymatosa.

Cases of phlegmonous inflammation of the oral cavity are rarer than those of the catarrhal form.

The cause is most frequently injury, as, for example, from foreign bodies, cauterisation, and sloughing as a result of swallowing caustic alkalies and acids. Erysipelas of the face may spread to the mucous membrane of the mouth, and produce vesicles, as well as phlegmonous swelling. Diffuse phlegmonous stomatitis is sometimes an independent disease, sometimes secondary to severe fevers, such as typhoid and scarlatina. It has its seat in the submucous connective tissue of the lips and cheeks.

The first symptoms are high fever, pain in speaking and eating, increased salivation, and exceedingly tender brawny hardness of the affected part. Later on, in the course of the disease, there are found here and there abscesses of varying size, or diffuse purulent infiltrations, which open into the mouth, and may lead to death from pyæmia or septicæmia.

The tongue is relatively most frequently attacked by inflammations of a phlegmonous type, which penetrate deeply into the tissue, the so-called "parenchymatous inflammations."

The tongue also swells rapidly when injured, as for example by pricks from foreign bodies, the stings of flies, wasps, or bees, or from infection from anthrax, but sometimes, too, without any demonstrable cause. Sometimes only one-half of the organ is affected, which fact leads some observers (Guéneau de Mussy and Dyce Duckworth) to attribute the affection to nervous influences. The swelling is due sometimes to œdema, sometimes to inflammatory infiltration of the parenchyma, which may be re-absorbed, or lead to suppuration. The symptoms of acute phlegmonous glossitis consist in the great and rapidly increasing enlargement of one or both halves of the tongue, which becomes too large for the mouth, and protrudes from between the lips, or presses against the back wall of the throat and larynx, thus causing asphyxia. These symptoms are generally accompanied by fever, intense pains, increasing to such a degree that partaking of food is rendered impossible, very profuse salivation, and fœtor. In the circumscribed form hard infiltrations often arise, which may be sooner or later re-absorbed, but generally develop into abscesses. Most frequently these occur on the base of the tongue, in front of the epiglottis, and on the side or tip.

Macroglossia, or *prolapsus linguæ*, is to be regarded as a chronic parenchymatous inflammation of the tongue.

It is sometimes hereditary, sometimes acquired, and consists in hypertrophy of all the tissues which constitute the tongue. Generally it is bilateral, though, according to Maas, it may affect only one side. We must ascribe acquired macroglossia to frequent inflammation or injury of the tongue, *e.g.*, bites, which are common in epileptic patients. The truly frightful enlargement of the tongue is the most prominent symptom, and causes it to protrude. It is forced out between the teeth, and on account of its dry condition is covered on its superior surface with fissures. The alveolar edges of the teeth, on account of the constant pressure of the hypertrophied organ, are forced outwards in a horizontal direction, the gum loosens, and the teeth fall out. Copious salivation, pain in eating and speaking, and an indistinct and stammering utterance are also constant symptoms.

On the boundary line between neoplasm and chronic inflammation we find *glossitis syphilitica indurativa*.

In the mucous membrane of the tongue, as also in its interstitial connective tissue and muscles, there is a proliferation of cells and increase of connective tissue, thus causing the tongue to be enlarged sometimes in one place, sometimes in another, and sometimes in its entirety. The organ is hard to the touch, and, in the circumscribed form, nodular, but absolutely painless. After some time the connective tissue shrivels up, and thus the tongue, being deprived of nourishment, appears rough and shrunken on its superior surface. This resembles, therefore, the process we see in other organs, as, for example, in the liver (hepatitis syphilitica or cirrhosis).

Among the chronic parenchymatous inflammatory processes of the oral cavity must be also reckoned *myxadenitis labialis* or *cheilitis glandularis apostematosa*, described by Volkmann.

It manifests itself by considerable swelling of the under lip, the mucous glands of which appear enlarged and develop into small abscesses. The swelling of the lip is developed gradually and without special pain; it becomes hard, firm, and often so bulky that it is quite immoveable. At the same time the mucous glands become as large as a millet seed, and rough, the excretory duct is widely dilated, and a muco-purulent secretion exudes on pressure. Boils are generally formed, which sometimes burst into the mouth, sometimes open externally, and often for months excrete muco-purulent secretion. In these cases we cannot trace a connection with syphilis. Volkmann recommends the following treatment:—Internal administration of iodide of potassium, antiseptic washes (*e.g.*, chlorate of potassium), and slight cauterisation of the lip.

Treatment.—The treatment of acute phlegmonous inflammation of the mouth and of glossitis must be energetic and antiphlogistic,—*e.g.*, ice compresses, application of pieces of ice to the tongue, and painting with diluted tincture of iodine (1:8). Deep scarification of the tongue longitudinally should be employed in threatening suffocation, and will usually be followed by speedy improvement. Should this treatment not relieve dyspnœa, tracheotomy must be performed. If it be difficult to administer medicines internally (quinine, for example, in high fever, or where the symptoms are of a pyæmic nature), it must be attempted per clysma, and in like manner action of the bowels must be procured. Should the diffuse swelling decrease and symptoms of abscess appear, warm washes must be ordered, and when fluctuation becomes marked, the abscess should be opened as soon as possible. Afterwards antiseptics are to be recommended, such as chlorate of potassium, carbolic, boracic, and salicylic acids. In macroglossia excision of a wedge-shaped portion, horizontally and vertically, is very useful, as also is ignipuncture, by means of which Helferich accomplished an almost complete cure in one case. Compression by means of bandages is hardly of any use, as their fixation is a matter of great difficulty. In glossitis syphilitica iodide of potassium should never be omitted.

EXUDATIVE STOMATITIS.

Synonyms.—Stomatitis Exudativa, Phlyctænulosa, Vesiculosa, Aphthosa.

In the oral cavity numerous changes take place, which are not very easily classified, either with the forms already described or with those to be discussed afterwards, but which on account of their frequency must be here considered. Under the above title it appears to me allowable to treat of herpes, pemphigus, variola, and aphthæ; against this classification no argument grounded on pathology can be adduced. Vesicles and blisters containing serum occur principally in herpes and pemphigus; pustules are observed chiefly in variola.

Herpetic eruptions are specially prone to attack the lips—*herpes labialis*—and the hard palate, more rarely the cheeks or tongue.

Etiologically, rheumatic and febrile conditions are most favourable to this disease; its appearance often heralds acute nasal catarrh, influenza, and muscular or articular rheumatism. How frequently it appears in pneumonia, and its almost constant absence in cases of typhoid, are well known. It has been maintained, but not proved, that herpes may be brought on by kissing or drinking out of one glass, although the person that has used the glass first may not have been suffering from the disease. Irritants, which affect the lips and mouth, especially food highly salted or peppered, and strong cigars, may, without doubt, promote its outbreak. Herpes labialis appears with particular preference just where the skin meets the mucous membrane of the lips. Sometimes on the upper lip, sometimes on the under lip, and sometimes on both, there suddenly appear very rapidly several blisters, generally in groups, varying in size from a pin-head to a pea, which, at first clear and transparent, become in a few days quite dim, and give the eruption a greyish white or greyish blue colour; sometimes they are filled with a purulent substance. The vesicles shrivel up in a few days, either of their own accord, or they may be ruptured by accident, especially when blowing the nose. Round the vesicles is a red ring due to inflammatory infiltration of the tissues. Should herpes be developed on the mucous membrane itself, it is very seldom possible to see the characteristic vesicles, because the elevated epithelial coating is macerated and shed in a short time. There are generally small round superficial excoriations perceptible, standing separately or in groups, which have a yellowish base and red or swollen edges, on which are visible the remains of the elevated epithelial coating.

While labial herpes usually only excites a sensation of heat and tension or stringing when touched, herpes of the mucous membranes causes considerable pain, which is still more increased by attempts to partake of nourishment.

The affection generally lasts a few days, but it may take a week or two till all the symptoms disappear.

Larger vesicles often follow erysipelas, which progresses from the external skin or from the throat to the oral cavity, or they may be due to pemphigus, but they are also excited by other causes, which as yet are not satisfactorily known. This is a fitting opportunity to mention a most remarkable case, described by Sidlo, of acute development of vesicles and emphysema in the mouth and on the soft palate, which was probably due to infection from septic or putrid material.

In rare cases, glanders, as well as foot and mouth disease, are localised on the mucous membrane of the mouth and throat. On the latter, according to Bollinger, there suddenly appear, along with febrile symptoms, headache, heat and dryness of the mouth, vesicles on the lips and tongue, more rarely on the soft and hard palate. These vesicles develop to the size of a pea, and disappear in a few days, leaving behind them dark red superficial ulcers and erosions. The common symptoms are pain in the mouth on masticating and speaking, diffuse redness of the mucous membrane, and difficulty in swallowing when the throat is affected. In glanders there appear, at a later stage of the disease, either simultaneously with or after the eruption in the nose, ulcerations on the gum, which bleed easily, emit a fœtid odour, and are followed by swelling of the submaxillary glands.

When variola develops on the mucous membrane of the mouth, the primary nodules develop into pustules, which burst and generate ulcers. The accompanying stomatitis is generally very violent, and the lymphatic and salivary glands swell with resulting salivation, when a considerable number of pustules are present.

Another affection belonging to this class is *Aphtha.*

While formerly the word aphtha was used to designate the most diverse conditions, the term is now limited to the occurrence of small flat whitish or cream-coloured spots or erosions, which have a favourable course, and never contain fœtid secretion.

Opinions still differ as to the nature of aphtha. While some ascribe it to vesicles, others to suppurating mucous follicles, or cell infiltration of the submucous connective tissue, Bohn traces it to croupous exudation between the submucous and epithelial strata. I have observed the condition repeatedly myself in adults proceeding from small hæmorrhages in the mucous membrane.

At different parts of the oral cavity, on the lips, cheeks, hard and soft palate, and especially on the tongue, there are formed spots as large as a pin-head or lentil, consisting of white masses, which are distinctly raised above the mucous membrane, and which appear suddenly in crops. The white coating is attached firmly at first to the underlying structures, is surrounded by a dark red livid border, and can be removed only by force, followed by bleeding; later, the exudation is softened by copious secretion, and is finally raised in consequence of mechanical influences. Although a considerable mass of epithelium remains, Bohn calls this " open aphtha," and protests against the designation of " ulcers," because under the latter term is understood purulent disintegration of tissue, and in many cases of mild and medium aphtha there is never any shedding of epithelium. When several neighbouring aphthous spots coalesce, as on the tongue for example, there are formed extended tortuous map-like spots or erosions, which may give rise to many diagnostic errors.

Although one may designate changes brought about by bleeding of the capillaries as hæmorrhagic erosions, and therefor as a special form of ulcer, the fact remains, that they do not differ at all from aphtha in their appearance and progress. The process of repair proceeds in the following manner: the aphthous mass is gradually limited and shed, and then the spots, being freed from exudation and epithelium, quickly heal.

Whether aphtha is indeed infectious, as many authors maintain, must appear doubtful. It is true that it is specially common in females, during menstruation, pregnancy, and the puerperal period. Many women suffer at each menstruation from this troublesome complaint; also the influence of dentition, the partaking of milk from cows suffering from foot and mouth disease, as well as of gastric disorder, on its development, cannot be denied.

The subjective symptoms of aphtha consist in pain and great sensitiveness of the mouth on partaking of food, masticating, speaking, or smoking. Infants and even older children refuse food, cry, and may become badly nourished.

The differential diagnosis of one kind of blister and vesicle from the other, is not usually very difficult. Herpes is most easily confounded with aphtha. While herpetic vesicles shrivel up quickly on the lips and form scabs, aphtha does not dry up at all. Then, again, on the one hand, herpetic excoriations are shallow, and appear as yellowish and furry erosions; on the other hand, aphtha is distinguished by greater depth and a white, fatty, thick coating. Both herpes and aphtha are distinguished from syphilitic, tubercular, and traumatic ulcerations, by their sudden beginning, which is usually accompanied by fever, as also by their rapid progress.

Treatment.—In treating all forms of exudative stomatitis, it is necessary to pay special attention to nutrition; simple foods, such as lukewarm or coldish milk, unsalted bouillon, and soup or eggs, should be preferred, on account of the great sensitiveness of the mouth. Of course alcohol and tobacco must be strictly forbidden. Since almost all mouth washes rather increase than allay the pain, local treatment should be limited to cleansing first with lukewarm water or camomile tea, perhaps with the addition of a few drops of tincture of opium, or rinsing with boracic acid (grs. 5-15 ad ℥i.). Painting with a solution of borax or with glycerine of tannin (grs. 15 ad ℥i.) may be likewise recommended. In cases of aphtha, excited by affection of the stomach, bitters are recommended, or in some circumstances mild laxatives, and painting with a solution of nitrate of silver or ether.

ULCERATIVE STOMATITIS.

Synonym.—Stomatitis Ulcerosa.

If, together with reddening and swelling of the mucous membrane, the epithelium is shed, and there is superficial and deep destruction of tissue, the condition is designated ulcerative stomatitis or stomatitis ulcerosa. Although every erosion may not be an ulcer, the latter is constantly developed from the former, and if on this account several affections associated with shedding of epithelium were discussed in the chapter on Stomatitis Exudativa, which are reckoned by others as ulcerative processes, I find myself justified in doing so, when regarding the matter from a pathological point of view. For the same reason I have thought it right to devote a special chapter to syphilis and tubercle of the mouth. As cases of ulcerative stomatitis, the so-called specific[1] mouth inflammations are considered in the first place. These, indeed, are not always combined with ulcerations, but frequently end in such.

Amongst these inflammations, *mercurial stomatitis* takes practically the first place.

Should mercury be lodged in the system for a very long time or in very large doses, it matters not whether through the stomach, the skin, the subcutaneous cellular tissue, or through the lung in form of mercurial vapours, mercurial poisoning first makes itself known by an unpleasant metallic taste and dryness in the mouth. Chemists, hat manufacturers, mirror liners, and barometer makers, are especially exposed to mercurial vapours. The teeth appear lengthened, the laying hold of and masticating food is painful, and is often followed by bleeding of the gum. Soon copious secretion and salivation set in, the pain increases, the tongue swells, and as it has now no room behind the teeth, it is protruded out of the mouth.

On objective examination the mucous membrane of the mouth appears *in toto* reddened and inflamed, but to a specially high degree the gum of the lower incisors, which is swollen and covered with extravasations.

The gum, instead of closely surrounding the teeth, retreats more or less from them, causing them to become loose. The tongue and mucous membrane of the cheeks are invested with a greasy, dirty, greenish covering, which emits an intense fœtor. If the affection be not cured at this stage, the mucous membrane becomes invested with a whitish grey membranous coating, on the removal of which are exposed irregular deep ulcers, which bleed easily. Salivation is most copious; the taking of food becomes always more painful, and at last impossible, the teeth fall out, while the ulcerations penetrate deeper and deeper as far as the periosteum of the jaw, and may thus produce periostitis, caries, and necrosis, or even gangrene. Fortunately such cases are very rare in the present day.

The diagnosis of mercurial stomatitis is dependent on being able to demonstrate the absorption of mercury or its salts. Easy as this

---

[1] Specific not used here in the sense of syphilis.

generally is, it is sometimes very difficult to arrive at a conclusion; and to assure ourselves of the difficulty sometimes met with, we require only to be reminded of an observation by Bamberger. A case of obstinate stomatitis ulcerativa appeared to him to be ætiologically inexplicable, till he at last traced it to mercurial ointment, which had been put on the bedstead to extirpate bugs. The author also is acquainted with the case of a doctor, who, in a few hours after coating the zinc elements of an electric apparatus with amalgam, was seized with stomatitis. Why one person should be sensitive to the smallest doses of mercury, and another stand the strongest inunction cures without any symptoms of poisoning, is very remarkable and difficult to explain.

The prognosis of mercurial stomatitis is, as long as it does not come to ulceration, a favourable one; it is, however, unfavourable or doubtful if extensive, or indeed even gangrenous ulcers are present.

Treatment.—The treatment must in the first place be directed towards placing the patient beyond the reach of injurious influences; all medicines containing mercury, and all substances impregnated with it, must be immediately laid aside. The dwelling or room in which the inunctions had up till then been practised must also be quitted. The local treatment, except in cases where ulcers are present, is identical with that prescribed for catarrhal stomatitis; but should the stage of ulceration have been reached, disinfectants must be employed in preference, as, for example, washes containing chlorate of potassium, boracic acid, aq. chlori, thymol, or extr. lign. campech., or a camphor emulsion (1 per cent.). Painting the ulcers with tannic acid (grs. 50 ad ʒi.), or with tinct. rhatan. (ʒi.), acid. carbol. (grs. 5), is also very useful. Many, like Vogel, prefer the internal use of chlorate of potassium, 4 grammes (1 gramme = 15·43 grs.) a day, both as a preventive and remedy.

Next to be considered is *stomatitis scorbutica*, which is one of the many localisations of scurvy, that chronic and general disturbance of nutrition which is manifested principally by hæmorrhages in the most diverse organs, and by rapid decrease of strength.

This disease begins, after having been announced by pain in the joints and general depression, with hyperæsthesia of the oral cavity, pain on mastication, and increased secretion of saliva. The gum, which appears encircled by a bluish red edge, swells, its processes between the teeth grow to knob-like cones, rise over their insertion, and are covered with larger or smaller extravasations. Gradually ulcers are formed, the edges of which are irregular and beset with fungous excrescences, and the base of which appears dirty, brown, or covered with fresh blood. The necrotic mucous membrane, which often hangs down in shreds, is covered by a diphtheric exudation, and bleeds at the slightest touch. The excretion is extremely profuse, the saliva mixed with bloody mucus, and extremely fœtid. The teeth sometimes fall out, sometimes become only very loose, and on the cheeks, tongue, and lips are developed hæmorrhagic effusions, which in turn change into ulcers. The hæmorrhages are often so copious that the patients in a comparatively short time sink from loss of blood.

The diagnosis of scorbutic stomatitis is not difficult, if attention be paid to other symptoms. In no other form of stomatitis is the swelling of the mucous membrane so intense. Stomacace, which is to be discussed immediately, is purely a local affection, and is characterised principally by yellow lines on the gums, which are absent in scurvy.

The prognosis of oral scurvy is in itself not unfavourable, but it is at least doubtful when the disease has reached a high stage of development, or when complications are present in other vital organs. Sudden copious hæmorrhages may also considerably darken the previously favourable prognosis.

Treatment.—In treatment of oral scurvy, the diet must be regulated, and unhealthy conditions of barracks, prisons, or ships improved. As defective quality or quantity of food (*e.g.*, absence of fatty constituents, indigestible nature, &c., or want of variety) are causes of scurvy, great care must be taken to provide changes of diet, and to see that its constituent parts are suitable. Formerly, fresh fruit, fresh vegetables, *saurkraut*, and water-cresses, as well as beer, alcohol, and citric acid, were considered good as preventives against scurvy. In cases of pronounced scurvy, Vogel prefers, before any other remedy, brewers' yeast—$\bar{3}$5-6 *per diem*. In very sensitive conditions of the oral cavity, decoctions of cinchona, gentian, and calam. arom., are principally to be recommended, while some speak highly of creosote and aqua picis.

The local treatment must also be considered, and the oral cavity most diligently cleansed by means of the washes, &c. mentioned in former chapters. When there is evident inclination to bleeding, the ulcers may be painted with diluted liq. ferri perchlor., or dusted over with alum.

Idiopathic stomacace bears great resemblance to the two forms just discussed.

Its etiology is not yet fully explained, but so far it is certain that the following causes play an important part in its ætiology—damp and overcrowded dwellings, insufficient and improper food, the influence of warm seasons, malaria, rickets, scrofula, and tuberculosis. Although stomacace is chiefly a disease of childhood, adults are not altogether exempt from it, as is proved by the numerous epidemics occurring among the military of France, Belgium, and Portugal. As its existence depends on the presence of gums—*i.e.*, teeth—it does not affect toothless children or old people. On the other hand, it visits with special preference unhealthy foundling hospitals, homes for sick, hospitals, and barracks. It appears sometimes sporadically, sometimes as an epidemic, and is therefore reckoned contagious, especially since Bergeron succeeded in inoculating himself. Those who maintain that it is not contagious support their argument on the negative results of some attempted inoculations, as well as on the fact that the officers and under-officers of the infected barracks were scarcely, if at all, affected by the disease. They believe that the spreading of the epidemic was less due to the contagion than to the soil and locality. Bohn considers that there is no doubt that stomacace is neither a disease of specific origin nor due to infection. According to this author, it declares itself by a local and chronic inflammation, and loosening of the gums from the teeth, excited and favoured by general causes.

Idiopathic stomacace begins generally acutely, seldom in a chronic manner, with symptoms of catarrhal stomatitis of the edge of the gum of the under jaw. In a short time—two or three days—or, in the chronic form, in weeks or even months, the swollen and retreating gum softens and changes to a yellowish, greasy, and putrid mass, after the shedding of which an ulcer is exposed, which spreads along the edge of the gum, bleeds at the slightest touch, and finally causes the teeth to fall out. Also, on the parts of the cheeks and tongue opposite the gum, there are formed, out of bluish red swellings, deep ulcers, the bases of which are covered with greyish yellow sanguineous pus. Swelling of the submaxillary lymphatic glands is also present in most cases. In the greasy material investing the ulcerations, besides blood and pus corpuscles, amorphous masses are found, and a number of fungi (especially *Leptothrix buccalis, Spirochæton,* and *Monas lens*). The disorder reaches its greatest severity when the periosteum of the jaw becomes affected, or when gangrene makes its appearance. The general health is, in most cases, little disturbed, fever is absent or very slight, the appetite often only impaired, but pain is nearly always very severe, and the odour from the mouth most offensive.

The prognosis of stomacace is generally favourable, especially if the patient comes under treatment early. The ulcers usually heal without scars, though such may remain, to the hindrance of physiological processes, in cases of great loss of substance, from gangrene or other causes. Necrosis of the jaw may, but rarely does, occur.

The diagnosis depends on the symptoms, the ætiology, and the endemic character of the disease; yet it is very difficult at the beginning to distinguish idiopathic stomacace from mercurialism or scurvy. In the latter disease, the gum is generally much more intensely swollen, of a bluish-red, livid colour, projecting over the teeth, while the yellow border is absent.

Treatment.—In the therapeutics of stomacace chlorate of potassium is the principal agent. Its external application as a wash is often prevented by the extreme sensitiveness of the oral cavity, as well as the tender age of the patient: to a child of four years is given internally 30 grains, and to an adult 75 grains in the course of a day. According to Vogel, the fœtor disappears in twelve hours, the gums stop bleeding, and, in short, improvement is everywhere noticeable. According to Bohn, recovery in some cases only takes place after extraction of the teeth from the diseased gum. In anæmic and delicate patients nourishment and fresh air must be most strictly attended to; then, when the patient has been brought through the worst stage with fluid nourishment, pulpy and by degrees solid food is indicated, also strong wines, beer, eggs, and quinine. As long, however, as any traces of ulceration exist, the oral cavity must be cleansed after each meal, and rinsed out with disinfecting aromatic waters; in the case of children it should be wiped out.

Having considered specific inflammation of the mouth, we now turn to more circumscribed ulcers, of which those that owe their existence to mechanical, chemical, or thermal irritation—the so-called *traumatic ulcers*—must be first considered, on account of their daily occurrence.

Best known are the ulcerations on the edges and tip of the tongue in consequence of sharp fragments of teeth, sharp stumps, or defective crowns, and also ulcers of the gum, from irritation caused by deposits of tartar. To this class also belong ulcers caused by foreign bodies, pieces of bone, needles, fish bones, or the imprudent use of strong acids and salts, as, for example, carbolic acid or boiling water. Next come the so-called dentition and hooping-cough ulcers of children; the former being caused by the cutting of the under incisor, the latter by the friction of the tongue against the corners of the teeth during the fits of coughing. Last to be mentioned are slits and ulcerations on the frenum linguæ, produced by the tongue being violently drawn out and pressed against the teeth during laryngoscopic examination.

The symptoms which such ulcers cause are generally limited to pain and sensitiveness in eating or speaking. More or less pronounced salivation and bleeding occur rarely, and are hardly ever of a serious nature.

Treatment.—The treatment of such cases is according to the cause; so, on the one hand, sharp teeth must be filed or extracted, on the other, foreign bodies or tartar removed. The same local applications as are used in catarrhal stomatitis recommend themselves, and, if necessary, touching with nitrate of silver.

Another kind of ulcer, formerly identified with aphtha, is the follicular ulceration on the hard palate of newly born children, called also *Bednar's aphtha*.

According to Bohn, there are found on the yellowish injected mucous membrane of the arch of the palate, or on and near the raphe, as well as on those parts of the hard palate situated near the upper alveolar processes, tubercles from the size of a pin point to a millet seed. They are of a white milky colour, and of a round shape, protruding more or less from the mucous membrane, and are called by him "mucous membrane milia" or "comedones;" by Epstein they are spoken of as "pearly epithelial accumulations of newly born children." According to Epstein, they are not gland follicles or milia, but result from the palatine laminæ not having joined regularly and at once, and leaving small cavities or spaces, which lead to the development of stratified epithelial nodules by the continued cell-proliferation of the walls.

Some of these tubercles remain unchanged for weeks, and even months, while others disintegrate into pus and form lamellar ulcerations, which are divided into halves by the raphe, and walled in by bluish swollen mucous membrane. The ulcers on the sides of the posterior corners of the hard palate—the plaques pterygoidiennes—are produced, according to Epstein, by lesion of the mucous membrane, which is tightly stretched over the lamellar process of the pterygoid. These ulcers are as large as a lentil, and of a round or oval shape; many of them are as sharply defined as if they had been cut out by a punch. They also run into one another, and may in cachectic children cover the whole mucous membrane, and even lay bare the bone.

The symptoms caused by the ulceration generally escape observation, as they only occur in infants, the attention of the relatives or the doctor being occasionally attracted by the child screaming, or being unable to suck. According to Bohn, they may form the starting-point of aphtha, diphtheria, and gangrene.

The diagnosis is founded on the stereotyped position, appearance, and erratic course of the malady. These ulcers should not be confounded with syphilitic affections.

Treatment.—Treatment is not necessary in spontaneous retrogression of the disease; on the whole, however, it is identical with that given in the case of aphtha.

Little known, and therefore almost always confounded with other processes, are *idiopathic ulcers* on the mucous membrane of the tongue.

By the term is meant superficial ulcers or excoriations, which cannot be traced to any dyscrasia, as, for example, syphilis. At present, in proportion as the literature on this subject is small, the names that have been proposed for this disease are numerous.

While some, following Caspary, designate them as "transient innocent plaques," others name them, with Unna, "circular spotted exfoliations of the tongue," or "map-tongue;" and others again, with Hack, call them "superficial excoriations," or follow Gautier in designating them "epithelial exfoliations of the tongue." It does not seem advisable to apply the term idiopathic mucous membrane plaques, as a quite different affection about to be described—viz., leukoplakia oris—has also received this designation.

Innocent superficial excoriations are found on the dorsum of the tongue. They have red, granular floors, and edges of a colour between sulphur and clay; the latter, i.e., the edge, is composed of epithelium and masses of fungus. The epithelium is never destroyed over any great area, which accounts for the slight symptoms and rapid healing of this affection. They are very erratic, and change their position and appearance almost every day. According to Gautier, their spreading is not caused by their extending themselves, but by coalescence with small neighbouring spots. He distinguishes the sharp-edged desquamation from that with twisted edges, the map-tongue proper. According to Unna, an exacerbation of the disease often precedes the commencement of menstruation, and disappears with its appearance. According to Schwimmer, it also seems often to precede an attack of leukoplakia.

Unna describes idiopathic erosion of the tongue as a disease of the epithelium, running a chronic course, but liable to acute exacerbations, and considers it due to nervous irritation. The German scientists deny every connection with syphilis, to which opinion I also hold; but Parrot believes the condition in question to be due to syphilis. Anæmia, gastric catarrh, and toothache are predisposing causes of innocent excoriations. According to Hack, heredity also plays a part. Gautier maintains that all causes that debilitate the system may give rise to the development of the malady, especially excesses—extreme fatigue, hysteria, and tuberculosis; and in children, gastric disorders and worms. Syphilis, according to Gautier, does not directly excite the malady, but only acts generally by weakening the system.

Its diagnosis is founded principally on its erratic, superficial, painless, and chronic course. Aphthous and herpetic ulcers show deeper loss of substance, great pain, and affect also other parts of the mucous membrane of the mouth, as, for example, the lips, palate, &c. Their progress, too, is rapid. Mucous patches are most commonly confounded with them, but they have a milk-white, bluish appearance, are preceded by infiltration, and do not rapidly change their appearance and position. Innocent excoriations are characterised by a yellowish colour at the edges and red floor, with no infiltration, and an extraordinary tendency to change in form, size, and position. The syphilitic patches also are found on the lips, cheeks, gum, and throat, and are very painful after desquamation of their epithelium, while the latter are confined to the tongue, and are almost completely painless.

The prognosis of these idiopathic ulcers of the tongue is so far favourable, that the affection has no tendency to assume a malignant character, but very unfavourable on account of its obstinacy and chronic course.

Treatment.—As regards therapeutics, opinions are very different. While some maintain the uselessness of any treatment, Unna shows very satisfactory results. He recommends stomatics, composed of flowers of sulphur, grs. 30; mixt. gummos. or emuls. amygd., ʒi.; also, hyposulphite of soda and glycerine, $\overline{aa}$, ʒss.; aq. dest., ʒi. But he found most effectual the following mixture :—

℞ Acid. sulphurosi
   Aq. menth. pip. $\overline{aa}$, ... ʒi.
   Flor. sulph. . . . . .
   Syr. simpl. $\overline{aa}$, ... ʒiss.
   Gum. tragacanth. . . . grs. 10.
Sig.—The mouth wash. Shake well.

After cleansing the oral cavity with tepid or coldish water, a mouthful of this mixture is to be held five minutes in the mouth, and forced into every corner of it. Generally three times daily is sufficient for this rinsing. Besides local, there must be general treatment in cases of irregularities of the blood, affections of the stomach or female genitals.

DIPHTHERITIC STOMATITIS.

Synonym.—Stomatitis Diphtheritica.

Diphtheritic stomatitis is to be considered as a further increase of the inflammatory process.

It generally occurs secondarily, extending from the throat to the oral cavity, and usually affects the hard palate, the posterior part of the tongue and the mucous membrane of the cheeks, and the gum near the last molars. Primary idiopathic oral diphtheria most readily attacks ulcers and fissures of long standing, or wounds consequent upon surgical operations. It very seldom affects the unabraded mucous membrane. I have observed only two cases of primary oral diphtheria—one without decided cause in an elderly agricultural labourer, who sank under septicæmic symptoms; the other in an apothecary, after the extraction of a tooth with an apparently infected key. The pseudo-membrane rarely attains the same extent in the oral cavity as in the throat; it represents a thin greasy covering, after the disappearance of which the mucosa, deprived of its epithelium, appears ecchymosed and tends to bleed readily. The change of diphtheritic ulcers into gangrenous ulcers may be found in the mouth as well as in the throat.

The prognosis of oral diphtheria depends upon the intensity and character of the malady, the general condition of the patient, and coincident complications in the nose and throat.

Treatment.—The treatment is the same as in diphtheria of the throat, to be considered later.

GANGRENOUS STOMATITIS.

Synonym.—Stomatitis Gangrænosa.

If we place gangrene of the oral cavity among inflammatory processes, this is so far justified by the fact that inflammatory symptoms always precede it, and it represents only one of the varied issues of inflammation.

Among the primary, *i.e.*, predisposing causes, the principal are, anomalies of the blood, poor nourishment, scrofula, scurvy, and severe infectious diseases, such as typhoid and scarlet fever, also parenchymatous exudations, and purulent infiltrations.

The greater the disturbances to circulation by inflammation through emboli and thromboses, and the more excessive the exudation of blood, pus, or fibrin, the greater is the danger of gangrene. In the oral cavity the gangrenous process may begin in the mucous membrane or in the deeper parts, and may be either circumscribed or diffuse.

If, in specially unfavourable circumstances, any of the ulcers described in the foregoing chapter may turn into gangrene, such an issue is always exceptional. This most readily occurs in diffuse phlegmonous, diphtheritic, mercurial, and scorbutic stomatitis, and in idiopathic stomacace.

Gangrene develops with peculiar preference, and usually without any preceding mouth affection, on the cheek. This form of gangrene has been designated Noma or "Water-cancer," and described as a distinct disease. But there are no grounds at all for this, as Bamberger and other authors rightly explain, since noma differs only from other kinds of gangrene in its position, but not at all in its nature. Moist gangrene is more common in the oral cavity, though the dry form also occurs.

The special etiological factors of noma are defective nourishment, and living in dull, damp, overcrowded rooms. Yet one should not, as Vogel points out, ascribe too great importance to the above-mentioned causes, otherwise the conclusion would be that this disease must be more common among the populations of large towns. It is certain that noma is commoner in low lying places, in maritime districts, and neighbourhoods subject to malaria (in Holland, for example), than in more elevated districts. It is likewise true that season is a matter of indifference, also that delicate children are subject to it as well as those weakened by any acute infectious disease, especially typhoid, measles, dysentery, or chronic intestinal catarrh; that females are particularly susceptible, especially between the ages of three and seven, and that healthy grown-up persons may also be attacked. Lastly, it is not to be denied that the outbreak of noma may be promoted by imprudent use of mercurial preparations in the case of weakly or decrepit persons while they are affected with serious illness, as gangrene may then follow mercurial stomatitis.

Noma either is secondary to already existing local diseases, as, for example, ulcerative stomatitis of the gum or cheek, or it begins apparently primarily as a circumscribed hard infiltration of the inner surface (varying in size from a cherry stone to a hazel nut) of one (commonly the left) cheek, near the angle of the mouth, also sometimes with the appearance of a pale red globule filled with muddy brownish yellow contents. Sometimes it is impossible to decide which of these two conditions was first present, since the case has generally already reached the ulcerative stage before treatment is sought. There is then discovered a superficial ulcer with discoloured floor on a hard infiltrated base, which can be felt on the external surface of the cheek, and is sensitive to pressure. The destruction within the oral cavity progresses daily, indeed hourly, so that in a few days the inner surface of the cheek, gum, and the half of one lip is changed into a fœtid, black, or blackish-brown mass. Simultaneously or a little later the skin of the cheek and the lower eyelid becomes swollen and œdematous, the surface has a fatty oily gloss and a bluish mottled colour, the lymphatic glands swell, the cheek appears sodden and hot, and in the centre there forms a bluish livid spot surrounded by a dark inflammatory circle, which in a short time shrivels into a brownish scab, and the removal of this scab exposes a deep gangrenous ulcer. If the cheek is once penetrated from both sides by the progressing disorder, the gangrene spreads in all directions without interruption to the lips, tongue, alæ nasi, eyelid, as far as the ear, forehead, and temples of the affected side, and downwards to the clavicle. The jaws, cartilage, and bones of the nose, as also parts of the cheek-bone, become necrosed, the teeth fall out, and the whole cheek sloughs away, so that the patient is changed in such a frightful manner as to be unrecognisable.

The most prominent among the subjective symptoms of noma in its later stage are apathy and deep depression; pain is often slight; fever absent at the commencement, or insignificant; the pulse is quickened; thirst increased; the skin cool or heated. The secretion of the oral cavity always becomes, as the disease progresses, more profuse, abounding in blood and sanious pus, and the odour becomes daily more offensive. Œdema of the lower extremities, lobular pneumonia, gangrene of the lungs or profuse diarrhœa, hastens exhaustion, so that death may occur in a few days. A fatal termination indeed often occurs before the local symptoms attain any special height, generally, however, in from four to eight days after perforation of the cheek.

The symptoms of the other forms of oral gangrene depend upon the primary disease from which it has developed in each case; the time of development also is very different, and depends on the character and the intensity of the inflammation. Pain, which was at first uncommonly violent, may be very trifling after the appearance of gangrene, or even vanish entirely. The secretion of saliva is increased, the breath offensive (cadaverous), and copious bleeding may follow erosion of obstructed vessels. The febrile symptoms, apparent at the commencement, give place to lassitude and depression, the heated skin becomes cool, the face loses its expression, rigors set in, and the patient sinks through metastatic inflammation of other organs due to pyæmia or septicæmia.

The prognosis of oral gangrene is in all forms very doubtful; even in a very circumscribed case under energetic treatment, and with the prospect of healing favourable, the process may suddenly assume a malignant character. The prognosis is always very bad in the diffuse variety, and in noma proper, especially when external circumstances are unfavourable, or when there are already complications in other organs. Progress towards recovery is announced by a cessation of the destructive process, and a distinct line of demarcation around the gangrene is formed; thus the necrosed mucous membrane is thrown off as stinking sloughs, or spongy masses, while on the floor of the ulcer laudable pus and granulations become developed. Healing never occurs without cicatrisation, and the severer the condition the greater are the subsequent functional disorders. The most noticeable of these are deformities of the face, synechiæ between the mucous membranes, narrowing of the mouth and nostrils, ectropion of the eyelids, and cicatricial lockjaw. According to Bohn, a relapse may occur in a case of noma progressing to recovery, or already healed, in which latter case the firm cicatrised tissue may also be destroyed.

Treatment.—As one generally cannot remove the causes that occasion or favour gangrenous stomatitis, the treatment must be directed towards preserving the strength, and towards attaining the most careful cleansing and hygiene. If sloughs are formed, they must be removed by scissors or knife, and an attempt made to restrict the further progress of the disease. Mineral acids, chloride of iron, and the actual cautery were formerly used for this purpose, but now the nitrate of silver pencil is rightly preferred. The pencil should be slightly pointed and methodically pushed from one part of the slough to the other, till it comes against resistant tissue. This manipulation is to be repeated daily till the gangrene is limited or ceases. Dusting the gangrenous parts with pulverised charcoal is one of the simplest means of removing the bad odour. In a case mentioned by Lange the gangrene healed remarkably quickly after the application of lint saturated in turpentine, aud changed frequently. Whether or not modern antiseptics have shown better results has not yet been proved satisfactorily. However, recent applications are preferable for cleansing and deodorising ulcers; we now use chlorate and permanganate of potash, salicylic and carbolic acids, iodoform and peroxide of hydrogen, in preference to acetate of aluminum, alum, and chloride of lime.

In nourishing and strengthening the patient many difficulties are generally met with. Those difficulties are caused by the aversion to food or medicines, and by the continual loss through the perforation of part of the nourishment which may have been taken with great difficulty. Feeding by means of the œsophageal tube or per rectum is also difficult in children. Decoction of cinchona, camphor emulsion internally, clysters of quinine and chloral, or the subcutaneous injection of morphia are usually indicated. The immediate removal of the patient to pure country air is, according to Bohn, the most effectual treatment, but unfortunately circumstances too often prevent this.

SYPHILIS.

Syphilis is particularly prone to attack the mouth.

Syphilitic mucous papules are most commonly observed, if we exclude cases of primary contagion from infected substances, tobacco pipes, cigar-tips, or unclean kisses. There appear at first on the mucous membrane red circular spots as large as lentils or peas, which are slightly prominent. Later the epithelium becomes milky white or like mother of pearl. If this epithelium be thrown off, the efflorescence appears deep red. If the papillæ or mucous follicles go on growing on the floor of the ulcer, the surface is uneven, and covered with pointed excrescences and molecular detritus. If the plaques are situated, for example, on the lips, partly on the mucous membrane, and partly on the cutis, that portion which is on the cutis shrivels to a blackish brown scab, while that on the mucous membrane retains its grey-white colour. If several papules coalesce, extensive ulcers result, which are deceptively like diphtheria. On the angles of the mouth the papules readily assume a longitudinal form, thus causing lineal or spiral fissures of the mucous membrane (the so-called Rhagades).

The favourite seats of syphilitic mucous patches are the lips, angles of the mouth, and the edges of the tongue, and next in frequency the mucous membrane of the cheeks, the gum, and hard and soft palates.

The progress of syphilitic plaques is slow and chronic; they disappear sometimes spontaneously, sometimes very rapidly after treatment. They may return, however, again and again after intervals of years, and belong to the most stubborn manifestations of a specific constitutional taint. The symptoms caused by mucous patches consist of pain in speaking, smoking, and masticating; fissures at the angle of the mouth, prevent it from opening to its full extent; and, as in every other inflammatory condition of the oral cavity, the salivary secretion is increased, but a fœtid odour from the mouth occurs only in exceptional cases.

In the later stages of syphilis, tubercular and gummatous neoplasms, which become ulcers when disintegrated, may be seen. Gummata develop comparatively seldom in the mucous membrane of the cheeks, and when this happens they are situated close to the angle of the mouth, or on the lip, most rarely on the floor of the oral cavity (Fischer), which, according to Langenbeck, always escapes. Without doubt syphilomata of the tongue are most common; they proceed from the mucous membrane, the muscular substance, and the interstitial connective tissue; and vary in size from a pea to a walnut. The anterior part of the tongue, especially at the edges and centre, is most commonly affected.

Gummata of the tongue develop almost always without pain, and cause very little pain even at the stage of ulceration. Such ulcers are generally sharply defined, irregular in shape, and have thickened edges.

They occur most frequently on the hard palate. Round or oval ulcers, affecting the entire mucous membrane and even bone, commonly situated in the middle line, but sometimes to one or other side of it, result from the breaking down of syphilitic nodules of the mucosa, or from gummatous ostitis and periostitis; eventually this condition may lead to perforation into the nasal cavity, speaking through the nose, and the regurgitation of food.

The diagnosis of syphilitic mucous patches is founded on the above described objective conditions, and the simultaneous existence of other symptoms of syphilis. Syphilitic papules are oftenest confounded with aphtha and herpes. While the latter are painful at the commencement, syphilitic plaques give rise to pain only after some time. Then, again, herpes begins with the formation of a vesicle; the syphilitic plaques as red spots, the size of a lentil or pea, which soon become invested by a pearly white epithelial covering. There is yet one more point of difference, viz., the aphthous and herpetic ulcerations heal from the periphery to the centre; the syphilitic papules from the centre to the periphery.

The diagnosis of ulcerating gummata of the oral cavity is sometimes very difficult, especially when they appear as the only manifestation of syphilis contracted long before, and which was believed to be entirely cured. The most certain diagnosis can be made when perforating ulcers of the hard palate are present; besides, in syphilis they only occur in malignant neoplasms, carcinoma, and lupus, the latter, however, having more the character of neoplasms. Co-existing cicatrices on the skin and in the pharynx, defects of the epiglottis and nasal septum, and loss of substance of the frontal bone, all point with certainty to syphilis. The differential diagnosis between gumma and carcinoma of the tongue presents the greatest difficulty. The following may be taken as the principal differences. While carcinoma, often from the very beginning, and always during its course, is accompanied by the most violent pain, gummata give rise to little or no pain; disintegration in carcinoma begins on the surface and proceeds inwards, in gummata from within outwards. While in carcinoma swelling of the glands occurs very early, and is very pronounced, in gummata it is altogether absent, or, if present, is but slight; with gummata also there are no sebaceous-like masses, which in carcinoma can be squeezed out. In some cases the diagnosis can only be made by examining pieces of the ulcers under the microscope, or by trying the effect of drugs. Now and then gummata are met with, which certainly, even after several weeks of treatment with iodide of potash, neither disappear nor improve; carcinoma, on the other hand, is not in the least degree influenced by such treatment.

The prognosis of syphilitic affections of the mouth is favourable, although mucous patches have a great tendency to relapse, and sometimes after gummata are healed, irreparable defects, distorting cicatrices, adhesions, and perforations remain.

Treatment.—In treating oral syphilis, the first thing to be considered is the stage at which the disease has arrived. If mucous patches are to be treated, a course of mercury is necessary, only in those patients, however, who have not been subjected to a course, or at any rate to a thorough course, or in those the subject of relapses, who for some time previously have not been treated anti-specifically, or, finally, in cases where the oral affection is very extensive, and is associated with manifestations of syphilis elsewhere. The author has designedly formulated the general treatment very precisely, because it is still very common malpractice to order a fresh course of mercury whenever a single isolated patch reappears.

Since in many patients recovery is very much retarded by the use of tobacco and alcohol, their discontinuance is a *sine qua non*. The manner in which mercury is to be administered depends upon each special case; one must always insist, however, on the greatest care being taken of the mouth. Touching the patches with solid nitrate of silver generally brings about their cure after a few days, without general treatment. Mouth washes, containing chlorate of potash, tincture of rhatany, tannic or boracic acids, are also recommended.

Iodide of potassium is the specific remedy for gummatous affections. Adults begin with 1 gramme (15·4 grains), children half a gramme, slowly increased to 2 grammes daily. In some obstinate cases it may be alternated with mercury or with Zittmann's decoction (Decoct. sarsæ co.). For local treatment, painting the ulcers with Mandl's solution (R Pot. iod. grs. 30, iod. pur. grs. 5, glycerine ʒi., acid. carbol. grs. 5) is recommended, or dusting them over with iodoform. When the edges of ulcers become hypertrophied, they should be touched with solid caustic, while papillomatous excrescences should be cut off.

TUBERCULOSIS.

Tuberculosis of the mouth occurs both primarily and secondarily. It generally begins in the soft palate; the posterior parts of the mouth, and especially the cheeks and hard palate, are usually affected.

The tongue is often the only part involved. On it tuberculosis appears in two forms, viz., circumscribed tumours, and diffuse miliary tubercles.

In the case of the former, Nedopil maintains that immediately under the mucous membrane, or in the deeper parts of the tongue, are formed multiple tubercles as large as peas, or even hazel nuts, which gradually force their way towards the surface, and which are found particularly at the edges, dorsum, and base of the tongue, but also on its under surface and near the frenum. The mucous membrane over these tubercles ulcerates sooner or later, and there remain characteristically formed ulcers, with slit-like openings and slightly gaping edges, which are often covered with little white nodules. On separating the edges, one sees that the destruction of the deeper parts is very much greater than of the superficial.

Of more frequent occurrence, however, is disseminated tuberculosis. On the edges of the tongue, and on other parts of the mouth in persons affected with pulmonary phthisis, but also occasionally occurring as a primary manifestation of phthisis, are seen prominent nodules as large as a pin-head scattered about or in groups, which, through the scarcely altered epithelium, appear of a greyish yellow colour. As the disease advances, the epithelium peels off from the miliary tubercles, the cheesy contents become removed, and there remain behind superficial ulcers of the size of a lentil, or more diffuse losses of tissue, which are covered with yellowish, often fœtid, thin pus, and frequently enclose small bright or pale red granulations.

These ulcers during their whole course retain their atonic character; they increase by new miliary eruptions breaking out at the edges; they never penetrate deep down, and have either no, or only a slightly infiltrated base, while the lymphatic glands in the neighbourhood are hardly ever affected to any great extent.

As regards the etiology, it seems to me that the opinion held by many, that tubercular ulcers arise from injuries to the tongue, *i.e.*, that ulcers originally purely traumatic change in time to tubercular, is quite untenable. It is true, however, that in past years men have been oftener affected than women. Why tubercle sometimes first attacks the mouth is quite as unexplained as is its occurrence primarily in the throat, larynx, or testicle.

The course of tubercle of the mouth is very varied. While the miliary form virtually never heals or comes to a standstill, and after weeks or months causes death, with phthisical symptoms in other parts, the caseous forms run a very chronic course. Sometimes the ulcers resulting from it heal entirely, leaving behind them deep scars; sometimes only partially and for a time. Many remain stationary, getting better and worse alternately. In most cases of this kind the disease attacks the lungs, brain, or stomach, and ends fatally.

The most important symptoms of tuberculosis of the mouth are the functional disturbances. While in the caseous variety there is scarcely any pain, and only a certain amount of difficulty in moving the tongue during speaking and mastication, in the miliary form the ulcers cause very severe pain, and give rise to alterations in taste, to fever and fœtor of the breath.

The diagnosis of disseminated miliary tuberculosis is not difficult, on account of the typical appearance of the ulcers, particularly if other organs, and especially the lungs, share in the disease. Should the latter, as well as other parts, remain unaffected, the disease may be mistaken for innocent superficial excoriation of the tongue, or for traumatic, aphthous, or herpetic ulcers. From the first, however, the tubercular ulcers differ, in that they run a very chronic course, and exhibit little or no tendency to heal; from traumatic ulcers, in that they appear on parts where friction of the teeth is impossible; and from the last, by the absence of a red inflammatory ring, but principally by the presence of miliary tubercles on the edges of the ulcers, and by their very chronic course.

Tubercular tumours of the tongue are, however, most often confounded with syphilis and epithelioma. While the gummata and losses of substance which accompany the former yield in time to iodide of potassium, tubercular ulcers do not show the least reaction to that drug; and while epithelioma of the tongue goes on giving rise to lancinating pains, and a high degree of swelling of the lymphatic glands in the neighbourhood, tubercle causes little or no pain, and but little or no lymphadenitis.

Examination by the microscope affords the surest means of diagnosis, showing in epithelioma characteristic cancer cells, while in the case of tubercle one sees giant cells and tubercle bacilli.

The diagnosis of syphilis combined with tubercle is very difficult. The author, who has for a long time had frequent opportunities of observing many such cases, agrees with Nedopil that at the beginning of the disease an anti-syphilitic treatment is of decided utility, but quite useless at a later stage, and that the ulcers always take on more of a tubercular character.

Treatment.—The treatment of tubercle of the mouth must be general as well as local. But as we cannot here discuss the therapeutics of phthisis, we must turn our attention only to local treatment.

Disinfectants must first be noticed:—(1.) In the form of mouth washes, such as chlorate of potash, boracic acid, biborate of soda, carbolic acid (1 in 200), thymol (1 in 2000); (2.) of pigments, such as a solution of creosote in glycerine (creosote min. 6, glycerine ʒi., spir. vin. ʒiv.). Tannic and carbolic acids in glycerine (tannic acid grs. 16, glycerine ʒi., carbolic acid min. 1½). Cauterising with solid nitrate of silver, scraping with the sharp spoon, the thermo-cautery, or galvano-caustic point, are good for undermined ulcers. Dusting with iodoform best removes any fœtor, next to that, weak solutions of permanganate of potash; circumscribed cheesy deposits on the tongue, as well as primary tubercular ulcers, must as soon as possible be removed with the scissors, or destroyed with the galvano-caustic point, all the more as their presence is a constant source of danger to the whole system.

PARASITIC DISEASES.

Among the diseases of the mouth, those caused by parasites take a peculiar and almost unique position.

The best known is Soor or Thrush. It occurs most usually in children, but also, though much seldomer, in adults. It takes the form of white patches and membranes, which are due to the copious proliferation of the *Oidium albicans* or soor fungus.

The ways and means by which this disease is spread are very various. It is propagated either by germs in the air, during parturition, by the child coming in contact with the fungus-covered vagina, or, and this is the most common way, by means of objects to which the fungus is sticking.

The fungus attacks chiefly stratified epithelium, the soft laminated layers being specially adapted for its deposition, so that its favourite localities are the mouth, throat, aditus laryngis, œsophagus as far as its cardiac end; also the vulva, vagina, and anal opening. It seldom attacks mucous membranes covered with cylindrical epithelium, though it may in exceptional cases attack the nose, larynx, trachea, and bronchi.

The age of childhood seems particularly predisposed to thrush. The reasons for this seem to lie partly in the numerous anti-hygienic conditions of the nursery, and partly in the greater amount of rest which the mouth has during that period of life, these favouring the attachment and settling of the fungus; but, above all, in the habit of nourishing children with slightly fermenting milk, or with saccharine and amylaceous substances. Children reared at the breast with proper cleanliness are very seldom attacked, while those artificially fed show the greater number of cases. Another reason is the insufficient care of the mouth, whereby the refuse of the food and the acid fermentation of the ingesta cause great uneasiness. I need scarcely add that the principal exciter and propagator of thrush is that which unfortunately is still used by so many to soothe children, namely, the disgusting sucking-bag (*vulgo*, *Schnuller*), filled with bread, milk, and sugar. Delicate constitution in children must also be looked upon as a predisposing cause, since it is observed that quite healthy children are very seldom attacked. It is not necessary that the mucous membrane of the mouth be in a catarrhal state. Although that condition supplies the fungus with better nourishment, it is only necessary that the fungus on its invasion finds fermentable material on which it can settle and grow.

In adults the disease occurs after severe illnesses which affect the blood, and the nutrition of the body, especially after or during typhoid, tuberculosis, pneumonia, diabetes mellitus, cancer, &c. The exciting causes here also are the reception and decomposition of easily fermenting substances, deficient amount of mastication, and careless cleansing of the mouth, which, however, may be excused somewhat in this case on account of the patient's extreme weakness.

On account of the easy transmission of the thrush fungus, its occurrence endemically in children's, foundling, and maternity hospitals is not to be wondered at.

Let us now take up the objective symptoms of thrush in children. The first symptom is the appearance of small whitish specks and membranes, of the size of millet seeds, scattered about, or grouped closely together, on it may be normal, but most usually on already inflamed, mucous membrane. The tongue and lips are always the first to be attacked. Soon the specks increase in size, either by the formation of new deposits, or by the coalescing of several neighbouring patches, so that in a few days large areas, or indeed the whole surface of the mucous membrane, may be covered with an adhering whitish mass. These thrush membranes have a rough granular surface, and form a firm tough layer, which sometimes takes on a yellowish, brownish, or even blackish colour, derived from the ingesta or from some colouring matter entering the mouth. On trying to remove one of the specks, it is accomplished with some difficulty in spite of the superficial seat of the fungus, and also causes a little bleeding; and whether they are removed artificially, or become detached in the course of the disease, they are renewed with wonderful rapidity. The mucous membrane below is more or less reddened, swollen, and sensitive. In rare cases shedding of the epithelium occurs, followed by erosions and superficial ulcers.

Of the subjective symptoms, the chief is pain on sucking or eating. Children refuse the bottle, nor do they take the breast any better; and if they attempt to do so, drop it again at once, crying pitiably.

Adults complain of a burning sensation and tenderness while taking food, these being ascribed more to stomatitis, present either before the thrush made its appearance, or set up by the thrush. If there be no stomatitis, or only a very slight degree, the affection may exist for several days without being noticed. Hence, in order that it may not be neglected, the mouths of infants and of young children should always be carefully examined. As a rule, gastro-intestinal catarrh is associated with this local affection of the mouth, which is either a complication occurring casually before the outbreak of the thrush, or is directly occasioned by the swallowing of sour saliva, mixed with the thrush fungus. If there is severe diarrhœa, the anus becomes reddened and excoriated, and the fungus may also be seen there. Very great danger is caused if the fungus spreads to the throat and œsophagus; in the case of the latter it may proliferate so much as to close the gullet completely. Taking food becomes more and more difficult, and at last impossible, all that is swallowed being immediately vomited. If the fungus spreads to the larynx, it gives rise to hoarseness; in some cases to breathlessness. Buhl and Virchow hold that bronchitis and pneumonia may be brought on by inhaling the fungus spores. Zenker observed a case in which the fungus was found in the brain, but such cases are of the greatest rarity.

The prognosis of thrush in children is, to say the least, somewhat doubtful. Even in the mildest cases it will interfere with the nutrition of the children for some time; but as a rule, when it occurs without diarrhœa in children on the breast, it disappears in a few days. When it occurs along with some other disease, it may cause the latter to become very much worse. The younger the child, the more severe the intestinal catarrh, and the more extended the mouth affection, the more unfavourable is the prognosis. When the throat, œsophagus, larynx, or lungs are attacked, the prognosis is very bad. The disorder may suddenly disappear by the shedding of large cohesive membranes, but the danger of their growing again is always present.

Thrush in adults must be looked upon as a very unfavourable symptom, as it only occurs in most critical conditions of the bodily strength. The prognosis, comparatively speaking, is best in typhoid and pneumonia, bad in diabetes, tuberculosis, &c.

The diagnosis depends, on the one hand, on the objective and subjective symptoms just described, and on the other on the nature of the fungus.

If, with a suitable instrument, one removes a piece of the white patches and specks, one sees, on putting it under the microscope, not only epithelial cells and schizomycetes of different kinds, but also numerous filaments, which are unequally jointed, and have lateral branches and buds. The latter are separated from the chief filaments by notches and partitions. The filaments themselves vary in length, being either straight or bent. They are colourless, with dark, sharp edges, showing violet coloured cavities filled with granules. The ends of the filaments are rounded off, sometimes swollen like clubs, or covered with little bladder-like formations. Besides these, there are seen, close to the filaments, strongly refracting fruit capsules, either round or oval, being the so-called gonidiæ, very like yeast cells, having dark edges. They occur either singly or gathered together in groups.

It is quite possible to mistake thrush for aphtha, but the presence of the thallus filament, as well as the fact that the aphthous coating cannot be rubbed off, guards one from error. On very superficial observation it may also be mistaken for spots of milk left scattered about in the mouth. From the above-mentioned follicular ulcers of the mouth, thrush is distinguished by its extensive spreading, the former being characterised by their stereotyped situation on the hard palate.

Treatment.—The treatment of thrush is principally prophylactic. Foul, impure air, and the presence of easily fermenting and mouldy substances in the nursery, must be strictly forbidden.

Strictest attention must be paid to the cleansing of the mouth, which should be done after each meal. When thrush has broken out, the mouth must be still more carefully attended to, and after each act of vomiting, it, and especially the folds of the cheeks, should be thoroughly cleansed of everything with a linen cloth and clean water, and the fungus masses must be carefully and energetically rubbed off.

In the treatment of this affection alkalies are specially useful, solutions of carbonate of soda or potash, biborate of soda, 1 to 30 to 1 to 10, without any addition of syrup, which would quite counteract any benefit from the treatment. In severe cases solutions of nitrate of silver (grs. 2½—10 ad ʒi.) should be used, with which the mouth should be painted hourly, after cleansing the mucous membrane with cold water. The gastro-intestinal catarrh is to be treated with red wine, rice and barley soups, white of egg, &c. When thrush occurs in the œsophagus, vomiting must be excited, at first reflexly by means of the finger or a feather, afterwards by sulphate of copper or ipecacuanha.

Among the affections of the mouth caused by fungi we must consider the so-called *black tongue*, in so far anyhow as it has yet been observed.

It has of course nothing to do with the usual colourings caused by food or medicines taken into the mouth. As a rule, the dorsum, but sometimes the middle, sometimes the edges, base, or point of the tongue, are the seat of spots as black as ink, with a rough, shaggy surface, which slowly spread, and may spontaneously disappear again. Between the hypertrophied papillæ of the tongue arises a fungus which, according to Dessois, has no resemblance whatever to any other appearing in the mouth, and hence has been called by him the *Glossophyton*. The black colour is communicated to the epithelium by its absorbing the spores of the fungus. The black spots consist of fine hairs, which can easily be stripped off by a spatula, representing either branched or non-branched filaments, or spheroidal egg-like spores : whether these fungi are identical with *Aspergillus nigricans vel fumigatus*, occurring in the ear, must be proved by further examination.

The affection does not seem to give rise to any special disorders. It is usually the discoloration of the tongue, a feeling of dryness in the mouth, and the disturbance of the sense of taste, that causes the patient to consult the physician.

As in thrush, the treatment must be guided by the parasitic nature of the affection.

Should *Leptothrix buccalis* appear in the mouth in large quantity, it may give rise to the formation of soft white, or whitish yellow tubercles, which often are elevated above the surface. (This parasite, even when living, is quite harmless.) They usually attack the base of the tongue, especially the part between the circumvallate papillae and the epiglottis. The tonsils, too, are very liable to be attacked ;

Hering has given the affection the name, *Pharyngomykosis benigna vel leptothricica.*

More minute details will be given under diseases of the throat.

One other affection must be mentioned here, viz., the *Stomatomykosis sarcinæ* of Friedreich. Friedreich observed, in many wasting diseases, such as protracted typhoid and phthisis, on the mucous membrane of the mouth, especially that of the tongue and soft palate, hoar-frost-like growths and membranes, which were made up of very numerous small-celled sarcinous fungi. The affection, however, does not call for special attention, although, unless examined by the microscope, it may be mistaken for thrush.

HÆMORRHAGES.

Hæmorrhages under and on the mucous membrane come under observation in the most different forms of disease.

Effusions of blood under the mucous membrane appear as red or black spots, ecchymoses of different sizes, or as nodes and extensive blood cysts. More serious free effusions are not very common.

As ætiological factors, injuries come first, also teeth extractions and anomalies in the condition of the blood. In adults the principal is hæmophylia, then come others nearly related to it, viz., purpura rheumatica, morbus maculosus Werlhofii, scorbutus and diabetes. In children, follicular ulcerations of the hard palate, catarrh of the mouth, catarrh of the stomach, thrush, croup, diphtheria, and also injuries and abnormal tension of the arteries during whooping-cough, give rise to more or less serious hæmorrhages.

According to Ritter and Epstein, a temporary disposition to hæmorrhage occurs in newly born children, which manifests itself by numerous parenchymatous hæmorrhages in different parts, internal and external, and which diminishes with each week of the child's life. But this affection, as a rule, only occurs in foundling hospitals, and among children who are very weakly, syphilitic or anæmic.

Hæmorrhages in epithelioma of the tongue are very profuse, and are mainly due to the erosion of the trunk or some of the branches of the lingual artery.

The prognosis depends, on the one hand, on the severity of the hæmorrhage, and, on the other, on the origin; it is very unfavourable in hæmophylia and epithelioma, as a recurrence of the bleeding is almost certain to take place.

Treatment.—The treatment consists in the application of astringent mouth washes, or in cauterising with solid nitrate of silver or some other hæmostatic remedy, such as liquor ferri perchloridi. When an alveolus is bleeding it must be well plugged with wadding, soaked in a solution of perchloride of iron, and then compressed, while in severe arterial hæmorrhages the life can only be saved by twisting or ligaturing the principal artery of the affected organ, as, for example, the external maxillary or the lingual.

NEOPLASMS AND TUMOURS

Neoplasms and tumours occur pretty generally all over the mouth, but those which are most dangerous occur on the lips, gums, and tongue.

We have first to consider an affection which only claims notice here from an anatomical point of view, since in it the neoplasm only affects the epithelium of the mucous membrane. It is the little known and little thought of *Leucoplakia oris.*

Synonymous with it are *Psoriasis, Tylosis, Keratosis, Ichthyosis linguæ,* and *idiopathic mucous patches.*

Nedopil and Schwimmer deserve specially great praise for their clinical description of it.

By the term leucoplakia oris is understood an affection of the mouth, progressing chronically, which manifests itself by the presence on the tongue, cheeks, or lips, of white spots or striæ, of the size of a lentil or of a pea, and either sharply isolated or irregularly formed. Histologically, these are due, as before mentioned, to a thickening of the mucous membrane, caused by extensive proliferation of epithelial cells, while at the same time an abundant growth of small cells takes place around the vessels.

The beginning of the disease is generally overlooked. An erythematous stage often precedes it, while at the same time there appear on different parts of the mucous membrane of the mouth, red or brownish red circumscribed spots, which later on become white patches.

The tongue is most often attacked. At the commencement of the affection, isolated spots are seen of the size of a lentil or a pea, slightly raised above the level of the mucous membrane. These are of a bluish white, opalescent, dirty white, or sometimes yellowish colour, and have a smooth regular surface. Later on they gradually increase in thickness, are of a dull white or silvery colour, and present cicatricial contractions, with a depression in their centre and elevations at their periphery. In these patches cracks and fissures occur from the movements of the tongue, causing bleeding erosions and superficial ulcers. It generally takes several years for the spots to become thick and compact.

These spots are to be observed, as a rule, on the cheeks and lips at the same time. On the former they are either isolated or sharply defined from the surrounding parts, or are irregular and in the form of wide bluish striae with smooth surfaces. On the latter, and also on the tongue, they are seen as horny spots of varying thickness, as hard as cartilage, scattered over a great expanse of mucous membrane like white scales. Ulcers seldom occur on the lips, but on account of the constant friction occasioned by eating and speaking, obstinate erosions may result. After some time the tissues at the base and in the neighbourhood of the patches become infiltrated.

The subjective symptoms of leucoplakia consist in unpleasant, but seldom painful, sensations on taking food, especially if the food be tart, spicy, or sour, also on taking alcohol, and when smoking or speaking.

One of my patients, a singing-master, complained of difficulty in moving the tongue while speaking, but especially when pronouncing the letter R. The symptoms are generally more intense at first, while the patches are thin and the epithelium not yet hardened, though in later stages the pain becomes greater from the presence of fissures and cracks in the patches. Excessive salivation or alterations of taste are only observed when there is extensive superficial disease. In some cases papillary growths occur on or near the patches, and by their size give rise to considerable discomfort.

The etiology of leucoplakia is in some respects still very obscure. It is certain, however, that no definite connection exists between it and syphilis, although very often persons are seized with leucoplakia who at some former time have had syphilis. Some of the exciting causes which must specially be mentioned are anæmia, diabetes, disturbances of the digestive tract, and the influence of tobacco, whether smoked or chewed. Men well up in years are most often attacked, women seldom. Some of the latter were inveterate smokers, and in others the vulva was also affected.

Although the diagnosis of fully developed leucoplakia generally presents no difficulty, still experience shows that it is nearly always confounded with syphilitic mucous patches. In the differential diagnosis, the following must be specially noticed. While the colour of syphilitic patches is a bluish dirty white, that of the patches of leucoplakia appears after a time as pure white, and should their colour, on account of the coating of the tongue, appear dirty or yellowish white, that need only be removed to show plainly the pure white colour. Again, while leucoplakia never attacks the under surface of the tongue, syphilis often does, and while the patches in leucoplakia retain the same colour for a long time, and otherwise remain unchanged, those due to syphilis change their colour very soon, because they heal, and the mucous membrane regains its normal colour.

Further, while the surface of the syphilitic patches is often very uneven and soft, and appears thin and pitted, as if macerated, that of the patches in leucoplakia is generally smooth or distinctly warty, and in consequence of the excessive proliferation of epithelium, rough and thick, particularly in the later stages of the disease, when the epithelium has become horny; and lastly, while the syphilitic patches, after running their course, leave ulcers behind, in leucoplakia ulceration never occurs,—cracks or fissures being the most serious result. A superficial observer might confound this disease with aphtha or herpes, but the rapid course of these disorders ought to prevent such an error. Similarly it may be mistaken for thrush; but in adults thrush only appears as one of the many symptoms of severe wasting disease.

The course of leucoplakia is an extremely chronic one, lasting many years. It may come to a standstill, in many cases it resolves, while with exacerbations it may continue for a whole lifetime.

The prognosis as regards cure is unfavourable, as regards life generally favourable. Although in the majority of cases leucoplakia may exist a whole lifetime, without doing any harm to the rest of the system, or even be cured, yet there can be no doubt that it bears some relation to epithelioma of the tongue. Leucoplakia may directly change into epithelioma; but according to Nedopil, the cases in which such a transition takes place are not those in which the disease is worst. Epithelioma is usually developed from longish deep fissures or shallow epithelial ulcers, the edges of which become gradually indurated, bleed easily, and are painful, and after several months are followed by swelling of the lymphatic glands.

The course of epithelioma of the tongue excited in such a way differs in no respect from other epitheliomata. It cannot be maintained, however, that the change from leucoplakia to epithelioma is a necessary result.

Treatment.—Treatment is as good as useless, as up to this time we possess no remedy which is able to prevent the formation of new patches, or to cure those already formed. It is, of course, obvious that disorders of the stomach, anæmia, &c. must be treated with the greatest care, and that hot foods and smoking must be forbidden. To treat leucoplakia with anti-syphilitic remedies, as is often done, is quite useless, and often hurtful.

The local treatment consists in the loosening and removal of the often very thick epithelial covering, and the most successful remedies are alkaline mouth washes, especially carbonate of soda, grs. 15 to ℥i. of water. When the mouth is very sensitive, add some tincture of opium or morphia. Schwimmer found benefit, in some cases, from the use of ½ per cent. perchloride of mercury solution and 1 per cent. chromic acid solution. Hillairet recommends the application of a stronger solution of chromic acid (1 in 8) every three or four days, gradually increasing the interval. Cauterising with solid nitrate of silver, with acids or caustic alkalies, aggravates the disease; but, considering the possibility of the disease becoming epithelioma, excision or destroying the patches with the galvanic cautery seems quite justifiable. Should either the one or the other erosion take on an epitheliomatous action, extirpation must be proceeded with at once, but this in no way excludes a return of the disease.

The neoplasms which occur most frequently on the lips and cheeks are polypoidal degenerations of the mucous glands—*mucous cysts;* these are most usually seen on the inner surface of the lips and on the frenulum of the upper lip. There occur also angiomata, papillomata, and epitheliomata, the last nearly always on the under lip.

Of the neoplasms on the gums, *Epulis* is the best known. Histologically its structure is partly a proliferation of granulations, partly fibrous and sarcomatous, and partly epitheliomatous. Of common occurrence are tooth cysts and odontomes. Arising from the periosteum of the jaws, and projecting more or less into the mouth, the following tumours may occur—fibromata, myxomata, enchondromata, osteomata, sarcomata, and carcinomata.

Among the retention swellings to be first noticed are those observed by Roth, due to the obstruction of the glands between the tongue and the epiglottis, whose contents become greasy plugs, and which, when they burst, may give rise to fœtor of the breath. On the expulsion of the contents the symptoms disappear.

The tumours of the tongue specially deserving mention are mucous cysts, angiomata, papillomata, lipomata, adenomata, fibromata, amyloid tumours (Ziegler), fibro-myomata, enchondromata, and hydatids.

Epithelioma deserves a prominent place in the consideration of malignant neoplasms of the tongue. As it occurs much more frequently in men than in women, tobacco and alcohol are specially set forward as exciting causes. It has been proved, indeed, that constant irritation of the mouth may lead to epithelioma, and that it may also be developed from ulcers and erosions; the manner in which it develops from leucoplakia has already been shown.

Its development takes place in two ways. It originates either as a primary node on or under the mucous membrane, which penetrates deeply into the muscular substance, and after a while softens and decays, or from the base and edges of an originally benign ulcer becoming indurated, breaking up, and infecting an ever-increasing extent of the organ.

The proliferation of epithelium usually begins in the mucous membrane, and forms proliferating centres and epithelial nests, out of which the so-called *epithelial plugs* can be pressed. Sometimes the surface of the ulcer is invested with papillomatous excrescences. Sympathetic swelling of the lymphatic glands occurs very early in the course of the disease. The disease may be limited to the tongue;

but experience shows that in most cases it spreads to the floor of the mouth, the palate, and even to the larynx.

The subjective symptoms of epithelioma of the tongue are,—so long as the tumour does not ulcerate,—a feeling of thickness and difficulty in moving the organ in speaking and eating. There may be no pain at first, but generally it occurs very early, becoming constant and exceedingly severe as the disease advances, and radiating towards the ear, jaws, or larynx. Speech becomes indistinct, stammering, the moving of the tongue daily more difficult, and the patient carefully avoids chewing hard substances. Salivation increases, the breath becomes fœtid, and sometimes severe hæmorrhages take place, endangering the patient's life. The strength fails, the appearance becomes pale and cachectic, the appetite diminishes, and in from one to three years the disease ends fatally, often with dropsical symptoms, or from metastatic pneumonia, or acute hæmorrhage.

For the differential diagnosis look back to page 33.

The prognosis is very bad. Even in the case of early excision, a recurrence generally takes place, particularly if a wide margin of apparently unaffected tissue be not taken away along with it. If a second operation is possible, it is certainly to be recommended; but, besides this, the only indications are, keeping up the strength, cleansing the mouth, stopping profuse hæmorrhage by means of ice, ligature, or an astringent preparation of iron, and, above all, full doses of some anodyne remedy.

As a conclusion to this chapter, may be mentioned the tumours on the floor of the mouth, which on the one hand consist of sebaceous and dermoid cysts, and on the other show themselves as retentive swellings or sublingual cysts.

The best known is *Ranula* (German, Fröschleingeschwulst). It generally occurs on one side only, but may be bilateral, as in a case lately demonstrated by Schäffer. Ranula occurs as a grayish white, fluctuating, elastic bladder-like tumour, varying in size, having its seat on the floor of the mouth under the tongue, but which, when more advanced, forces itself up between the tongue and lower jaw. There is generally no pain, but speaking and chewing are both considerably impeded. Up to the present time, opinions regarding the nature of ranula have been very divergent; while some ascribe it to an obstruction of Wharton's duct, and others regard it as a dropsy of one of the mucous follicles on the outer side of the genioglossus muscle; it is now certainly proved by Recklinghausen and Sonnenburg, that ranula proceeds from the so called Nuhn-Blandin gland, situated beneath the anterior part of the tongue.

The prognosis is so far unfavourable, in that the tendency is for it to recur.

Treatment.—The treatment must be operative. Should the swelling refill after simple incision, with or without the subsequent injection of iodine, then a wider opening must be made to empty out the contents, by removing a piece of the cyst wall with scissors; after which the walls of the cyst must be stitched to the mucous membrane with several sutures.

The treatment of the remaining neoplasms and tumours mentioned above, must be conducted on the lines laid down in text books of surgery.

DISEASES OF THE NERVES.

The various diseases of the mouth are in most cases accompanied by alterations of common sensibility and also of taste.

Diminution or complete loss of sensibility in the mouth, or anæsthesia, hardly ever occurs, except in diseases of the brain, or in peripheral paralysis of the second and third branches of the trigeminus. The affection is usually one-sided, as in localised lesions, and extends to the mucous membrane of one cheek, or to half of the tongue. In different lesions of the brain, as well as of the spinal cord, and, according to Romberg and Althaus, after the continued influence of cold air, the anæsthesia may become bilateral. When the tongue is attacked, disturbances of taste also take place.

When the anæsthesia is unilateral, food taken into the mouth is perceived on the affected side, only on account of its temperature or its physical properties. A glass put to the mouth gives the impression of being broken off, and remnants of food collect between the cheek and gum, and remain there unobserved. The tongue is thickly furred on the anæsthetic side, and on it are seen cracks and bites, inflicted by the patient himself, from his not knowing the position of his tongue, especially when the motor fibres of the nerve are also affected.

Hyperæsthesia of the mouth may arise from central lesions certainly, or from hysteria, but is as a rule caused by neuralgia of the trigeminus. It extends to the palate, cheeks, lips, and gums, and only in rare cases to the tongue, floor of the mouth, or anterior pillar of the fauces. The pain is generally of a neuralgic character, beginning suddenly, then disappearing after a few minutes, and returning again at stated intervals. Salivation is, as a rule, increased, and one half of the tongue furred. According to Albert, the neuralgic pains are often caused by little excrescences on the edge of the tongue, immediately in front of the base of the anterior pillar of the fauces.

The same etiological conditions, as a rule, also cause *paræsthesia* of the mouth. At one time the patient complains of a feeling of fur in the mouth, at another of "falling asleep" of the tip of the tongue, and of heat and cold in the gums, or hard palate. The author once observed in a case of acute spinal meningitis, altered sensation limited to the point of the tongue, which, however, disappeared gradually along with the other symptoms.

Occurring more frequently than anomalies of ordinary sensibility are disturbances of the sense of taste.

Anæsthesia of the nerves of taste (anæsthesia gustatoria, ageustia) may be produced in a purely mechanical way, by an impediment to the action of sapid substances on the peripheral nerve endings, as, for example, from abnormal dryness of the mucous membrane, from the tongue being thickly coated, by food either very hot or very cold, or, on the other hand, by central and peripheral lesions of nerve conduction in the trigeminus, lingual, chorda tympani, facial, and glossopharyngeal nerves. The power of taste may be either diminished or arrested, for all or only for some kinds of taste, the alteration being complete or partial, unilateral or bilateral. In the mildest cases, the perception of taste is always somewhat delayed. Lesions in the region of the trigeminus, lingual, chorda tympani, and facial, principally interfere with the power of taste in the anterior two-thirds of the tongue, but also in its tip and edges. There can be no doubt that affections of the glossopharyngeal are followed by loss of taste at the root of the tongue, the palate, and posterior wall of the pharynx. When the glossopharyngeal and lingual nerves are simultaneously affected, and there is also anæsthesia of the skin, the cause must be looked upon as being of central origin (Seeligmüller).

Hyperæsthesia gustatoria is often a physiological phenomenon. Just as there are persons with unusual sharpness of sight, or with unusually good sense of smell or hearing, so there are also those with extraordinary delicacy of taste, especially if the last has been trained from youth, as is often seen in the case of wine merchants, tea and coffee buyers, and gourmands, who are able to determine the vintage and variety of wine, the different sorts of tea, and the least taste of spices.

Pathological hyperæsthesia of taste occurs relatively oftenest in hysterical or nervous persons, who then can taste the minutest quantity of salt or spices, which healthy people do not perceive at all. The author once observed a case of periodical hyperæsthesia of taste in a healthy young man, to whom food, quite free from salt, appeared at these times excessively pungent and unpleasant.

Altered perception of taste very often occurs, and is known as *Paræsthesia gustatoria* or *Allotriageustia*. It occurs quite commonly in fevers, and in disorders of the alimentary canal, in short, in all these processes in which the tongue becomes furred and the mucous membrane of the mouth in any way altered. Patients with rheumatic facial paralysis also often complain of an insipid, sourish, or bitter taste. Favourite dishes even generally have an unpleasant metallic or foul taste. Sugar seems bitter, salt sweet, coffee tastes like chicory, cigars like burnt straw, &c. Here also must be mentioned the partiality which so many chlorotic, hysterical, or pregnant women have for unusual, distasteful, and even nauseous things, such as burnt coffee beans, hemp seed, chalk, onions, assafœtida, castor oil, &c. Many of these anomalies are to be considered as distinct hallucinations of taste and illusions, and looked on as manifestations of psychical disturbance.

Treatment.—The treatment of these different conditions promises results, only when careful consideration is given to the etiology; but since an exhaustive description of these extremely numerous affections would carry us far beyond the limits of this book, it is only possible to mention here the favourable influence of electricity, the constant current being more effectual than the induced. It is easy to understand that the disturbances arising from central lesions present scarcely any prospect of a good result.

Disturbances of movement manifest themselves as spasm and paralysis.

Spasm occurs principally in the tongue, and generally as part of some general nerve lesion, such as hysteria, eclampsia, epilepsy, chorea, etc., and is sometimes clonic, sometimes tonic in character.

Idiopathic spasm, limited to the tongue, is very rare, although several authors, in particular Berger, have observed it. The tongue is drawn intermittently forwards and backwards, it is turned and raised, or is thrown here and there in all directions with smacking noises, just as in the case of a person stammering. In tonic spasm, according to Valleix, the tongue is immovably fixed against the hard palate.

Spasm of the masseter occurs much more commonly. It is either tonic or clonic in character. The former, which is called *Trismus* or *lockjaw*, is generally bilateral. The lower jaw is so firmly pressed against the upper that the teeth cannot be separated, nor the mouth opened; the muscles are as hard as a board, and very painful. If only some of the muscles of mastication are affected, or only on one side, the lower jaw becomes displaced, as in a case of Leube's, in which the pterygoids on one side only were involved, the lower jaw on the affected side being caused to protrude beyond the upper.

In the clonic form the lower jaw is quickly jerked against the upper: if this be in a vertical direction, the teeth chatter; if horizontal, they grind on each other. It is easy to understand how these involuntary and violent movements may cause the tongue, lips, and cheeks to be bitten, and the gums and teeth to be more or less injured.

The causes of this affection are meningitis, cerebral tumours, epilepsy, hysteria, encephalitis, and more particularly the different forms of tetanus, tetany, and hydrophobia; it is also, but rarely, caused by peripheral irritation of the trigeminus consequent on bad teeth.

The prognosis, when the affection is due to central lesion, and also in tetanus, is in the highest degree unfavourable, but is favourable when the cause is peripheral.

Treatment.—The treatment in the first place must be directed to the removal of the exciting causes, but this is impossible in the majority of cases. In tetanus, the exhibition of full doses of narcotics is to be recommended, especially morphia subcutaneously, also the inhalation of chloroform, and warm baths. When rheumatism is the cause the patient should be made to perspire freely; salicylate of soda should be given, and galvanism applied to the muscles. When the tongue is affected, galvanism should be applied to the hypoglossal nerve, and also to the medulla oblongata.

Of far more importance than the nervous lesions which have just been considered, are the *paralyses* which attack the various organs composing the oral cavity.

Paralysis of the lips, especially of the orbicularis oris, is an almost constant accompaniment of central or peripheral paralysis of the facial nerve; it is generally unilateral, affects sometimes the lower lip, sometimes the upper, sometimes both, as is generally the case in bulbar paralysis. Articulation becomes indistinct on account of the partial or complete inability to pronounce the labial consonants. The patient is also unable to whistle, or to blow out a light, all the more if, as is so often the case, the muscles of the cheeks are likewise affected. If the paralysis is complete the mouth cannot shut, part of the food falls out, and the saliva continually dribbles over the lips and chin.

Paralysis of the masseter is generally of central origin, being a symptom of inflammatory, degenerative, or neoplastic disease of the pons or medulla oblongata, but chiefly of bulbar paralysis. The lesion may be, but is very seldom, caused by pressure on the nerve supplying the parts.

The earliest of the subjective symptoms is difficulty in masticating hard substances, powerlessness of the muscles being complained of by the patients. Objectively there is loss of tone or relaxation of the muscles, mastication becomes gradually more difficult, and at last quite impossible; the under jaw, when the lesion is bilateral, hangs down relaxed, the muscles atrophy, and sometimes become contracted.

Paralysis of the tongue manifests itself in a specially marked manner.

If the lesion be unilateral, as is usual in embolus and apoplexy, or in peripheral pressure on the hypoglossal, the tip of the tongue when protruded inclines to the paralysed side, the reason being that the non-paralysed genioglossus draws the tip of the tongue over the lower jaw towards the opposite side, because the corresponding muscle on the other side being paralysed offers no resistance. In advanced long-standing cases, atrophy of the paralysed half of the tongue takes place.

If the lesion be bilateral, the tongue cannot be protruded at all, sometimes not even as far as the teeth, but lies in the mouth like a dead atrophic mass, only showing now and again fibrillar contractions. If the lesion be incomplete (paresis), the point of the tongue cannot be protruded either so far or so quickly, nor can it be so promptly moved about in different directions, as it can be when in its normal condition.

The disorders caused by paralysis of the tongue bear, on the one hand, upon the function of mastication, on the other, upon articulation.

While, under normal conditions, pieces of food are pushed by the tongue between the two rows of teeth to be bitten small, and are then moved backwards into the throat, in paralysis of the tongue the food lies on the dorsum of the tongue and remains there, and even if it does get into the pharynx, it is simply pushed back into the mouth, because the wall formed by the tongue between the mouth and the throat is awanting, or is incomplete.

The disorders of articulation, in unilateral paralysis, show themselves in indistinct pronunciation of the linguals; in bilateral, and especially in bulbar paralysis, speech is stammering and almost unintelligible.

The prognosis of paralysis of the tongue depends upon the cause and on the degree of the lesion. In peripheral pressure on the hypoglossal, under certain conditions the paralysis may be removed by a surgical operation, extirpation of a tumour, or extraction of a retained bullet, or any other foreign body; in apoplexy or cerebral embolus the prognosis is not altogether so unfavourable as the paralysis may after a time get well; but in tumours of the brain encepnalic deposits, and bulbar paralysis, it is absolutely hopeless.

Treatment.—The treatment of paralysis of the tongue, lips, and muscles of mastication consists in the simultaneous treatment of their cause, in the electrical excitation of the hypoglossal nerve, or of the muscles of the tongue and of the cheeks from the oral cavity. In bulbar paralysis, the highest attainment is arrest of the disease; scarcely any improvement, or even an approach to recovery, can be recorded, although Tommasi by means of faradism, and Benedikt by galvanism of the sympathetics, consider that they have had good results.

DISEASES OF THE SALIVARY GLANDS.

The efferent ducts of the salivary glands all being situated in the mouth, it seems proper that their diseases be classified with those of the mouth.

There are three salivary glands, the parotid, the submaxillary, and the sublingual.

The largest and most important of these is the parotid. It lies in front of, and under the ear, in the recess bounded by the ascending ramus of the lower jaw, the mastoid process, and the external auditory canal. It extends from there across the external surface of the masseter, as far as the inferior edge of the zygomatic arch. Inwards it reaches to the styloid process. It is lobular in appearance, and consists of rounded segments called *acini*, bound together by connective tissue. Its efferent duct, known as Stenson's duct, is distinguished by the thickness of its wall and narrowness of its lumen, and is situated at the upper third of the anterior border of the gland. It is formed by the meeting of the ducts of the acini, runs parallel to the zygomatic arch, then under it forwards on the outer surface of the masseter, at whose anterior border it descends though the fatty layer of the cheek to the buccinator muscle, which it pierces at its middle point, to open on to the inner surface of the cheek, opposite the first or second upper molar tooth. In a living person the orifice may be seen, if the cheek be withdrawn, as a black speck, no bigger than a pin head, or as a small hollow, and into it a fine probe or canula can be passed. Sometimes the opening is covered with a small button-like protuberance. Sometimes there is seen lying in front of the parotid gland, or on its efferent duct, another smaller gland called the *Parotis accessoria*, whose duct runs into Stenson's duct, and round about the implantation of the latter, in the buccinator sometimes there is found a group of small buccal glands.

The salivary gland of the lower jaw, *i. e.*, the submaxillary gland, is about half the size of the parotid; it lies under the mylohyoid muscle, between the superficial and deep fascia of the neck, in the triangular space bounded by the inferior edge of the lower jaw and the two bellies of the digastric muscle. Its efferent duct, called Wharton's duct, along which a delicate process of glandular tissue extends as a prolongation of the gland substance, passes over the superior surface of the mylohyoid, between it and the sublingual gland inwards and forwards, and opens on the rounded extremity of one of the papillæ, which are to be found on both sides of the frenum of the tongue (caruncula sublingualis).

The sublingual gland is the smallest of the three, and lies on the upper surface of the mylohyoid, covered only by the mucous membrane of the floor of the mouth, which it elevates to a slight degree. Its posterior end lies in contact with the anterior end of the submaxillary gland, and its minute ducts, 8 to 12 in number (known as the ducts of Rivini), open either behind the caruncula sublingualis into the mouth, or run into one another, and, as in the case of the other glands, form a common duct. This is known as Bartholini's duct, and it as often has an opening of its own on the caruncula, as it unites with Wharton's duct. It is still undecided whether the so called Nuhn-Blandin gland, which lies under the point of the tongue, is to be considered as a salivary or as a mucous gland.

The saliva secreted by these glands has for its function the moistening and saturating of the mouth, and also of the ingesta. By its influence starch is converted into grape sugar.

Without describing the physiological details of the secretion of the saliva, let us consider a very frequent pathological phenomenon, viz., increased flow of saliva.

SALIVATION OR PTYALISM.

Increased salivation is the constant accompaniment of most of the diseases of the mouth. Particularly does it occur in children when cutting their teeth, also in mercurial stomatitis, aphtha, and in variola of the mouth. It also is caused by the examination of the mouth, throat, and larynx, as well as by taking irritating articles of food and certain medicines, principally mercury, iodine, copper, and lead; also during dental operations, when the saliva is poured out in a jerky manner. Not only does increased secretion of saliva occur in diseases of the mouth, but also in both acute and chronic diseases of the stomach, bowels, and uterus. It is caused reflexly by the presence of entozoa in the alimentary tract, and by anomalies of the blood, as in anæmia and chlorosis, also during pregnancy and diseases of the throat and nose. A very remarkable form is that occurring in hysterical persons of both sexes. It may occur in consequence of pain in the face, teeth, or ear, thus confirming the fact that ptyalism depends to a very great extent on the nervous system, and especially on irritation of the trigeminal, facial, and glossopharyngeal nerves.

The onset of salivation is frequently accompanied by an unpleasant dragging feeling, never by real pain, in the region of the parotid gland and lower jaw, and often by an insipid metallic taste. The quantity of saliva excreted is often prodigious, and may amount to from 2 to 9 pints in the day. Naturally the mouth cannot make use of such quantities, hence some of it is swallowed, giving rise, especially in children, to catarrh of the stomach and diarrhœa; the rest flows out of the mouth, compelling the patient to expectorate continually, and even depriving him of his night's rest.

When the skin is very sensitive, but especially in the case of children, reddening and even erosion of the skin of the neck and chin takes place; and according to Vogel, on account of the continual wetting of the clothing, and the chilling caused thereby, bronchial catarrh may be set up.

Another consequence of the increased secretion of saliva is a diminished secretion of urine; sometimes it leads to emaciation which is due to loss of appetite, caused by dilution of the gastric juice and the consequent limited digestion of starchy materials.

In colour the saliva is almost always turbid (milky) and opalescent, due to the admixture of epithelium with it, sometimes it is more viscid and stringy, sometimes thin and watery; in stomatitis it has a disagreeable, even fœtid odour.

The specific gravity of the saliva is at first considerably increased, even as high as 1059, but it gradually sinks below normal, and often only amounts to 1001. Its reaction varies: sometimes it is normal, sometimes alkaline or acid, or it may be neutral. The quantity of sulpho-cyanate of potash in it is usually considerably lessened. In mercurial ptyalism, Bamberger has, by means of galvanism, detected mercury in the saliva.

The duration of the affection varies very much, and lasts from a few days to several months, or even years, according to the primary disease. Nervous ptyalism is often very intermittent, occurring at longer or shorter intervals.

The prognosis is, as a rule, favourable, even when, as already mentioned, disturbances of nutrition are present on account of the long duration of the affection. Recovery seldom occurs suddenly, but is nearly always gradual.

Treatment.—The therapeutics of salivation can but rarely be prophylactic, as when it is caused by mercury or iodine. In administering calomel, which, as a laxative in acute diseases, is still much esteemed, caution must be exercised, as has already been mentioned in speaking of gangrene. It is self-evident that on the first appearance of salivation, the proper remedy must be at once and thoroughly administered. Diseases of the mouth, and of the alimentary canal, are to be suitably treated, and bad teeth extracted. The treatment of those cases in which the exciting causes cannot be made out is very difficult. Sometimes success is attained by exciting the secretions of the skin, the stomach, and the kidneys, by means of douching, steam baths, friction, and other methods of stimulating the action of the skin, as well as by mild laxatives and copious water drinking. Nervous ptyalism is most stubborn, so also is that occurring during pregnancy; it disappears most readily on the administration of opium, chloral, morphia, and bromide of potassium. Others have seen benefit follow the use of atropin, acetate of lead, and iodide of potash. Locally, astringent mouth washes of alum, tannin, tincture of rhatany, bi-carbonate of soda, etc., are to be recommended, especially where inflammation of the mucous membrane of the mouth is also present. Attention must of course be given to ordinary nutrition; starchy materials being limited as strictly as possible.

PAROTITIS.

There is no doubt that, of the salivary glands, the parotid is the one most frequently attacked by inflammation, although at the same time the submaxillary and sublingual are by no means exempted.

Parotitis may occur primarily or idiopathically, and secondarily or metastatically.

Affections of the mouth, injuries, foreign bodies, and concretions in Stenson's duct, and also catching cold, play a very subordinate part in the origin of idiopathic parotitis. Gruber holds that inflammation of the middle ear, which does not extend to the external auditory meatus, may, by way of the Glasserian fissure, cause parotitis and pharyngitis.

Idiopathic parotitis being usually an infectious disease, appears generally as an epidemic.

Although up to the present time we are still in the dark as to the nature of the poison, yet we must conclude from the different symptoms of the disease that it is due to infection. Epidemic parotitis is a purely contagious disease, and is usually spread by the inhalation of the specific poison coming from a person suffering from parotitis. It is, moreover, not merely a local affection, but a general infectious disease, primarily localised in the salivary glands. Children from 2 to 15 years old are specially liable to be attacked, but persons at the age of puberty also furnish a fair proportion of cases. Males are particularly predisposed to it, and it generally occurs during the cold seasons of the year. Sometimes it is limited to a very small space, to a boarding school, an orphan institution, a cadet corps, a barrack, or even to a single flat in a house; while, on the other hand, it may spread over an entire building, a whole village, a town, or even a province.

The spread of the disease, as a rule, goes on slowly. Incubation generally lasts from 7 to 14 days, though it has been observed to be only 3 or 4 days, and sometimes much longer, even from 20 to 21 days. Some physicians do not regard parotitis as infectious; if, however, we consider that other diseases also, as for example cholera, in spite of their introduction to a place, do not spread epidemically, because there are special circumstances necessary for their doing so, the often observed isolated cases of parotitis will not appear remarkable.

The anatomical basis of parotitis is due, on the one hand, to catarrh of the mucous membrane of the ducts of the gland, and also of the gland itself; on the other, to inflammation of the interacinous and periglandular connective tissue. Many observers, and among them Gerhardt, consider the term periparotitis, a more suitable name, since the function of the gland is not interfered with during the inflammation.

The disease begins, as in some epidemics, with a distinct prodromal stage, this being less pronounced in others, the symptoms being flying stinging pains in front of the ear, a feeling of tension when the mouth is opened, lassitude, want of appetite, sleeplessness, and slight fever.

After two or three days symptoms in the gland itself make their appearance; opening the mouth and chewing become painful, the parts in front of the ear are swollen, the swelling extending in all directions, so that the lobe of the ear and the rest of the auricle are raised and forced outwards, and the region in front of the external auditory meatus and the zygomatic process appear bulged out; the swelling further extends to the lip and cheek, and also to the eyelid and orbit, producing a comical expression of the face, hence the disease is called "*Tölpel*" (booby), "*Bauernwetzel*" (country fellow), mumps, &c. In some epidemics both glands are affected, in others only one. Other glands may participate in the swelling. The first of these are the submaxillary and sublingual, which become, as I can affirm, sometimes alone, or at least earlier, more severely affected than the parotid itself. Next come the cervical, jugular, and axillary, lymphatic glands, and lastly, as was first observed by Gerhardt, the spleen.

The swelling of the parotid is in most cases only superficial, the buccopharyngeal, and deep fascia of the neck, preventing its penetrating further, but it can extend more or less deeply, pushing aside the lateral wall of the pharynx, and the isthmus of the fauces, and thus causing difficulty in swallowing and dysphagia, also hoarseness and dyspnœa. Difficulty in swallowing is in most cases brought on by the function of the muscles of the neck being interfered with, in consequence of the presence of the enlarged glands, or it may be due to stomatitis. Patients anxiously avoid opening the mouth or chewing, hold their heads stiffly and erect, and complain of stinging pains, noise in the ears, and difficulty in hearing. The skin over the swelling is stretched, pale, œdematous, and shining. Fever is always present, in some cases slight, in others very pronounced. The type of the fever is, as a rule, irregular, but there are generally evening rises and morning remissions. According to Debize the temperature may be from 39.5 to 40.0 C. for several days, and in rare cases takes on the character of typhoid fever, accompanied by apathy, delirium, and torpor.

Besides the lymphatic glands and the spleen, the genital organs are particularly liable to be affected, more usually in males, and at the time of puberty, seldom in boys or in old men. This form of orchitis often attacks only one, and that usually the right testicle, seldom both. Sometimes only the testicle is affected, sometimes the epididymis, sometimes both, and with it acute hydrocele may be associated. This complication sets in accompanied by a considerable amount of fever, also by dragging pain and a feeling of tension in the scrotum and inguinal region. The orchitis generally disappears, but atrophy of the testicle and impotence may result, and now and again it is associated with inflammation of the spermatic cord, cystitis, and urethral blennorrhœa.

In females there occur inflammation of the ovaries, vulvo-vaginal catarrh, inflammatory swelling of the vulva and breasts, with considerable pain.

As to the significance of these complications there has always been divided opinion. While some recognise them as simple metastases, others regard them, and evidently with more reason, as secondary localisations of the specific poison of parotitis, just as bronchitis after measles, and as affections of the kidneys, joints, and throat in scarlet fever. As to localisation in the brain, there are no reliable observations.

The duration of the disease varies very much with the character of the epidemic. Mild cases generally recover in from 3 to 4 days. Severe cases, especially when the genital organs are implicated, last from 2 to 4 weeks. Still longer is the duration, if, as is frequently the case, a relapse takes place. The relapse may last for from 8 to 14 days more, rarely does it continue longer, though in a case observed by Gerhardt, it continued for 19 days after complete recovery from the first attack.

The prognosis is favourable, since the affection almost always results in resolution, very seldom in ulceration. In malignant epidemics life may be constantly in danger from the height of the fever, difficulty in breathing, and from extensive suppuration of the glands.

The diagnosis depends on the objective and subjective symptoms. If distortion of the face is already present, it is extremely easy. It may be confounded with swelling of the glands of the neck, but only when one is entirely ignorant of anatomy. The diagnosis is more difficult when the disease begins in the submaxillary gland, but the participation of the parotid later on puts one on the right track.

Treatment.—The treatment in mild cases may be expectant; in severe cases, especially with considerable amount of fever, quinine, salicylate of soda, kairin, or antipyrin should be given, and the bowels should be kept open, but not by means of calomel. From the use of emetics I have never found the least benefit. Locally, rubbing the swollen glands with clean fat or olive oil, or with vaseline, and bandaging with Brun's cotton wool, are recommended. Where, in the acute stage, inunction of a highly diluted grey ointment was used, I observed in most cases aggravation of the disease and salivation. The application of ice could not as a rule be tolerated. When recovery is slow, and where there is induration, a good application is tincture of iodine and tincture of galls in equal parts. Orchitis is to be treated by suitable position of the parts and cold applications, which generally quickly relieve the pain. When the female genitals are affected, syringing the vagina with lukewarm water, sitz baths, and warm poultices are most to be recommended; in dysphagia leeches should be applied, ice, cold gargles, or scarification of the œdematous parts; abscesses must be matured and opened soon.

Secondary or phlegmonous metastatic parotitis is a far more serious disease.

It generally follows close upon severe infectious diseases, particularly typhoid and scarlet fevers, but may also be met with after measles, variola, puerperal fever, pyæmia, erysipelas of the face, recurrent exanthematous fevers and typhus. Against its originating from catarrhal obstruction of Stenson's duct and consequent blocking up with putrefaction of the secretion, is brought the fact of its unusual rarity in catarrhal stomatitis, in thrush or in ulcerative stomatitis.

Parenchymatous degeneration of the glandular tissue caused by increase of temperature, must be taken as the exciting cause of this metastatic typhoid parotitis. According to Hoffman, typhoid parotitis, which has become undoubtedly less since the introduction of the cold water treatment, consists in an excessive increase of the changes in the parotid usually seen in typhoid, and bears the same relation to the latter, as the formation of ulcers and perforation of the intestine do to infiltration of the follicles. While in all severe cases of typhoid, the swelling of the glands, which is always present, and which is associated with degeneration of cells, usually resolves, without having made itself specially apparent; here, on isolated spots, it goes on to suppuration and disintegration of the tissues, not only of the gland substance itself, but also of the connective tissue between the acini. By the union of several small deposits, larger abscesses are formed, and in certain circumstances the whole gland may become converted into an abscess. Hoffmann explains that it is on account of the compactness and firmness of the fascia surrounding the parotid that violent inflammatory symptoms occur so much when the gland is swollen, while the other salivary glands and the pancreas are not attacked by the same process.

The subjective symptoms of suppurating parotitis are at first identical with those of the idiopathic form. On account of the usually severe prostration of the patient, and of the functional disturbance of the brain, complaints of pain in the ear, or of pain on opening the mouth, are not expressed. The swelling is generally very considerable, and as pus forms, the skin becomes reddened, fluctuation appears, and the abscess matures in a few days, but in protracted cases it may go on for weeks. There is usually a considerable rise of temperature. Disintegration is not localised to the parotid alone, but spreads into the surrounding parts, and leads to venous thrombosis, to suppuration of the masseters and pterygoids, and also to gangrene and pyaemia. The bursting of the pus takes place either towards the external auditory meatus, or through the skin over the parotid, and implication of the facial nerve often causes paralysis of several of the muscles of expression.

Parotitis suppurativa generally appears, according to Liebermeister, during the third or fourth week of typhoid, but almost only in severe cases, and must be looked upon as a very serious complication, because on the one hand it is a sure sign of the widespread degeneration of the organs, and on the other because it aggravates the diminution of strength by fresh fever, or threatens life by pyæmia.

The prognosis is the more unfavourable the earlier the affection appears in the course of infectious diseases. In prolonged or gangrenous disturbances, there remain disfiguring defects, caries of the bones, and paralysis of the face. *Ephidrosis parotidea*, a condition sometimes remaining after the healing of parotid abscesses, must be looked on as a most remarkable occurrence. During mastication a clear fluid is perspired from the skin over the parotid, while between meals none is observed.

Treatment.—The treatment consists in the earliest possible removal of pus. As Stenson's duct is sometimes stopped up, it is recommended to remove all hindrances and evacuate the pus, according to the method of Mosler and Bruns, by introducing canulæ, and also by making pressure on the parotid and clearing out the efferent ducts. Should this not succeed, poulticing, and then opening with the knife, must be resorted to. The incision must be made either outside the region of the blood vessels and the facial nerve, or if necessary, parallel with the latter, and through the parotid fascia; an antiseptic dressing is absolutely necessary. When patients are in a low state, special attention must be paid to strengthening diet, alcohol and quinine.

NEOPLASMS AND CONCRETIONS.

According to recent researches, simple hypertrophy of the parotid does not appear to be so common as was formerly supposed. It is occasioned by proliferation of all the constituent parts of the gland, or by fatty deposition or proliferation of the perialveolar and interacinous connective tissue. Idiopathic or secondary inflammations of the parotid are the usual causes of the resulting enlargement. I once saw in a man, aged thirty-eight, a soft, painless bilateral hypertrophy, giving rise to great distortion, but which could be attributed to no cause.

Cystic tumours of the salivary glands are fairly common. They may grow to the size of a hen's egg, and have partly serous, partly sebaceous contents.

Cystic tumours of the parotid are partly due to inflammatory ulceration of the mouth. They develop slowly and without pain, and occur on the cheek, or along the course of Stenson's duct; rarely in the parotid itself. They vary very much in size, seldom growing larger than a walnut. They are hard and firm, sometimes fluctuating. Cure takes place after reopening the efferent duct, or through disintegration and atrophy, and sometimes after abscess of the parotid and formation of fistulae.

Enchondromata are also of frequent occurrence: they spring from the stroma of the connective tissue of the glands, and are, according to Koenig, solitary nodes of real hyaline cartilage, which, growing from the deep parts of the glands, push aside the tissues, and appear on the surface as one or more lobulated masses. Numerous transition forms also occur, such as myxochondromata, myxosarcomata, and fibro-myxochondromata.

Carcinoma of the parotid is not uncommon, occurring in the form of scirrhus and medullary cancer in persons of advanced age. The former gives rise to tuberculated, uneven, hard swellings, which often take years to develop, chiefly on account of the fascia offering resistance to its growth; but when the latter is once penetrated, swelling of the lymphatic glands and ulceration set in. The soft cancer grows very rapidly, enlarges in all directions, forces its way into the mouth, and soon changes into suppurating ulcers.

The symptoms consist partly in distortion of the face, partly in interference with mastication, and also in pain, deafness, facial paralysis, compression of the external auditory meatus, increase of growth into the orbits, amaurosis, etc.

The prognosis is guided by the size and histological structure of the growth; innocent circumscribed tumours may leave behind incurable facial paralysis.

Treatment.—The treatment can only be operative. Previously, parenchymatous injection of iodine or chloride of zinc, or the internal use of iodine was tried, which in malignant lymphosarcoma was followed by decided benefit. Should these methods fail, then recourse must be had to extirpation of the parotid; for a description of which operation text-books on surgery must be referred to.

The parotid is also the seat of foreign bodies and concretions.

The former enter the gland itself or its efferent duct, either through the skin by means of injuries, or by the mouth.

As in the nose, so also in the salivary glands and their efferent ducts, little foreign concretions form, the so called *salivary calculi*, which are generally of the size of a pea, but may become as large as hazel or walnuts. They have a stratified structure, with alternate white and yellow layers, and consist for the most part of carbonates and phosphates, chalk and magnesia, with a small amount of organic material. Maas found numerous bacteria in them: their colour is usually clear white or yellow; sometimes they seem to be congenital.

The symptoms of concretions depend upon the blocking of the parotid duct. The saliva does not flow, the gland swells up and suppurates, and not infrequently salivary fistulæ are formed.

To diagnose the lesion, examination of the duct with a probe or small tube is indispensable.

Treatment.—The treatment consists in the dilatation of Stenson's or of Wharton's duct; if this does not bring about the desired result, then the concretions must be excised.

The submaxillary and sublingual glands are affected in the same way and at the same time as the parotid, or they may be affected singly. The submaxillary is more frequently attacked. There appears under the lower jaw a firm elastic tumour, accompanied by considerable pain and tension, and inability to open the mouth; it may become absorbed, or may go on to suppuration. If the inflammation of the submaxillary extends to the cellular tissue under the chin, then the disease is designated

ANGINA LUDOVICI, OR CYNANCHE CELLULARIS MALIGNA.

It appears spontaneously without apparent cause, after cold, or after a fall or blow on the under jaw; sometimes it follows an infectious disease, and must then be considered as metastatic. I saw such a case once in a child six months old, in whom, without doubt, the cause was suppuration of the sublingual gland.

The skin between the halves of the lower jaw, as far down as the hyoid bone, or even the sternum, is generally extensively infiltrated, and there are also well marked febrile and gastric symptoms. The skin is also reddened, and very painful, and the movement of the lower jaw is completely in abeyance. On account of the pressure of the ever-increasing swelling, the tongue is forced against the hard palate, so that speech, eating, and swallowing become impossible. In well marked cases stenosis of the larynx results from œdema of the mucous membrane, or from compression; and on account of pressure on the jugular veins, giddiness, tinnitus, and delirium may occur.

The course of the disease is generally protracted. The swelling may certainly resolve; this, however, seldom happens; usually abscesses form after a longer or shorter time, which are opened by the surgeon, or which burst of their own accord, and discharge, either into the mouth or externally, thick sanious pus and bits of dead cellular tissue. Fistulous openings, and scars, which interfere with the movement of the muscles of the neck, not infrequently remain.

The prognosis, regarding which I agree with Bamberger, is not altogether so unfavourable as is generally affirmed. That the metastatic form is worse than the idiopathic is certain, as also the fact that death may ensue in both forms from gangrenous destruction of the skin, and from pyæmia, or from asphyxia on account of the larynx becoming also affected.

Treatment.—The treatment at first should be antiphlogistic. In the idiopathic form local blood-letting is recommended, but not in the metastatic, being too weakening: ice too may be applied. Poultices are necessary to mature the pus, and incision must be made as soon as possible. For the fever, quinine must be given; for weakness, stimulants. When there is danger of choking, deeply incising the skin, or scarification of the mucous membrane must be tried, and eventually, if necessary, tracheotomy, which in these cases is a very difficult operation.

SECTION II.

DISEASES OF THE THROAT.

DISEASES OF THE THROAT.

ANATOMICAL AND CLINICAL INTRODUCTORY REMARKS.

By the terms, throat, pharyngeal cavity, and pharynx, is understood the elongated four-cornered space, which extends from the base of the occiput and sphenoid bones above, to the superior border of the fifth cervical vertebra below, and which is continuous anteriorly with the cavities of the nose and mouth, laterally with the middle ear, and inferiorly with the larynx and œsophagus. This space is lined by a fibrous trabecular framework, which, covered with mucous membrane, is attached to the anterior surface of the vertebral column, and contains numerous muscles, nerves, glands, and vessels. It is in relation with the vertebral column by means of the retro-pharyngeal cellular tissue, which extends over the aponeurosis covering the prevertebral muscles. On anatomical as well as clinical grounds, the pharynx is divided into several parts: the superior, extending from the base of the sphenoid and occiput to the uvula, is called the *pars nasalis*, or *pharyngo-nasal cavity;* the middle, extending from the uvula to the base of the tongue, is named the *pars oralis*, or *pharyngo-oral cavity;* and the inferior, extending from the base of the tongue, or from the point of the epiglottis to the opening of the œsophagus, is spoken of as the *pars laryngea*, or *pharyngo-laryngeal cavity.*

As it cannot be denied that this division is rather an arbitrary one, Rückert proposes that only two sections be recognised, sharply separated from one another by the isthmus of the fauces: an upper or air passage, and a lower, or food passage. While quite approving of this proposal, I have, on practical grounds, confined myself in this text-book to the old division.

The *pharyngo-nasal space*, or *naso-pharynx*, resembles, to a certain extent, a rounded cube; its superior wall, the roof of the pharynx, is formed by the body of the sphenoid and basilar portion of the occiput, and merges without sharp definition into the posterior wall, in the formation of which the atlas and the first cervical vertebra take part. The anterior wall is formed by the nasal fossæ, the vomer, the horizontal plates of the palate bones, the pterygoid plates of the sphenoid, and the posterior surface of the soft palate, which, by its contraction, sometimes goes to form the inferior limit of the nasopharynx. The lateral walls are formed by the fossæ of Rosenmüller, the openings of the Eustachian tubes, and their prominent margins.

The mucous membrane of this space is of some importance. On the roof, posterior, and lateral walls, to about the level of the openings of the tubes, it is specially rich in glandular organs. This cushion, rich in crypts and follicles, bears also the name *adenoid tissue*, or *pharyngeal tonsil*. It appears as a soft piece of tissue, elevated above the mucous membrane, which is divided into numerous prominences and elevations by narrow and wide slits, and is at its lower end provided with the *bursa pharyngea*, which has in the middle a small round opening of the size of a pin-head (foramen cœcum). This bursa lies behind the pharyngeal tonsil, extending as far as the body of the occiput; it often becomes constricted, and may, if its opening becomes closed, give rise to the formation of cysts.

The most important organs in the naso-pharynx are the Eustachian orifices and their prominent margins. By the projection of these margins there comes to be behind them a depression, known as the fossa of Rosenmüller, or *recessus pharyngis*, which contains both mucous follicles and adenoid tissue. The elevation has an anterior and a posterior lip, and between them is the opening of the tube. The latter varies in size, is sometimes funnel-shaped, sometimes like a slit, and is directed forwards and downwards. The anterior lip of the tube is also called the hook, and its continuation forms the hookfold or *plica salpingo-palatina*; the inferior half of this fold becomes flattened during phonation and swallowing.

Of great practical significance is the *plica salpingo-pharyngea*; it is very rich in glands, and takes its rise from the posterior end of the cartilage of the tube. It extends downwards from the plica salpingo-palatina parallel to the lateral wall of the naso-pharynx, and crosses the posterior border of the levator veli palati, which on its outer side extends obliquely down in front of it. Immediately above the posterior pillar of the fauces it becomes flattened, and extends backwards and outwards, turning towards the same pillar, on the superior surface of which it extends outwards. By its moving inwards and downwards during phonation and deglutition, it shuts off, according to Zaufal, the inferior from the superior cavity of the pharynx.

In the *pharyngo-oral cavity* the soft palate or velum palatinum is of special interest. It somewhat resembles a curtain, continued back from the hard palate, its two surfaces are covered with mucous membrane, and it is supplied with numerous muscles. Its anterior surface is continuous with the mucous membrane of the mouth, and its posterior with that of the naso-pharynx. The anterior surface is under ordinary circumstances slightly reddened and smooth, and contains the openings of numerous closely packed mucous glands, which appear as little elevations of the size of millet seeds, and are often covered with drops of mucus. In the middle is a cylindrical process, rich in glands, the *uvula*, while the inferior free edge divides into two diverging folds of mucous membrane, a posterior and smaller one going to form the posterior faucial pillar, and a wider and anterior one going to form the anterior faucial pillar, the former being inserted into the postero-lateral wall of the pharynx, and the latter into the lateral border of the tongue. Between the two pillars on either side, are the tonsils. The space bounded by the tonsils, the faucial pillars, the uvula, and the tongue, is called the *isthmus of the fauces*, and the two free spaces between the uvula and the pillars of the fauces, the *arcades*.

The *tonsils* are in adults oval glandular masses about the size of hazel nuts. They vary very much in size, form, and colour. It may even be affirmed that normal tonsils are scarcely ever seen in adults. On their free surface are numerous small roundish or slit-like openings, which lead into longer or shorter perpendicular or oblique slits, the lacunae or crypts. The lacunae are lined with a thin epithelial layer. Over the whole tonsil there is a firm connective tissue, between the fibres of which lies the gland substance, *i.e.* the follicles. There are no mucous glands in the tonsil.

The tissues surrounding the tonsils are of great importance. Anteriorly and posteriorly they are covered by loose, externally by firm, connective tissue, which is continuous with the bucco-pharyngeal fascia. The internal carotid lies one and a half centimetres, and the external carotid two centimetres, external to the tonsil.

The pharyngo-oral cavity has no anterior wall at its upper part; a space called the *aditus pharyngis* being anterior to it, and below this the base of the tongue forms its anterior boundary.

The *pharyngo-laryngeal cavity* is bounded anteriorly by the epiglottis and the aditus laryngis, laterally by two longish mucous cavities, the sinus pyriformes. The anterior surface of the vertebral column forms the posterior wall, both of this cavity and of the pharyngo-oral cavity, and is flat, slightly convex or slightly concave, or it may form a gentle arch from above downwards. In the pharyngo-oral cavity there are in most people, either in the middle line or at the sides, one or more yellowish red prominent spots, about half the size of a lentil, the so-called granulations, which will occupy our attention again later on. The mucous membrane of the posterior wall of the pharynx is slightly reddened, the seat of numerous vessels, is moist, and at its lower part somewhat granular.

The mucous membrane of the naso-pharynx is invested with ciliated, the rest of the pharynx with pavement epithelium.

The glands embedded in the mucous membrane are of two kinds, conglomerate and follicular: the former occupy chiefly the naso-pharynx, especially the posterior surface of the soft palate, the neighbourhood of the tubes and the fossæ of Rosenmüller; the follicular, round and circumscribed, are numerous in the naso-pharynx, especially on the pharyngeal tonsil, the Eustachian prominences, the fossæ of Rosenmüller, and also in the pharyngo-oral cavity.

The lymphatic vessels of the pharynx form a dense network in the mucous membrane; they are very numerous in the soft palate and in the tonsils, and also in the uvula. The glands in connection with them lie at the angle of the jaw, at the bifurcation of the carotid, inferior to and near the superior part of the sterno-mastoid, and at the side of the hyoid bone and larynx. The glands at the angle of the jaw communicate with the tonsils, so that it is not to be wondered at that they swell up when the latter are in any way affected.

The blood vessels of the pharynx are arterial and venous. Of the arteries, must be noticed the superior palatine from the internal maxillary, supplying the soft and hard palates; the ascending palatine from the external maxillary, supplying the mucous membrane, muscles, and glands of the palate; and the tonsillar artery, supplying the tonsils, lateral wall of the pharynx, and root of the tongue, also the ascending pharyngeal artery, the terminal branches of the internal maxillary and the vidian, and the pharyngea suprema to supply the naso-pharynx.

The veins form two plexûs: the posterior, which is in connection with the veins of the nasal mucous membrane, and joins the plexus in the temporal region; and the anterior, which communicates with the root of the tongue, and joins the internal jugular by means of the pharyngeal vein.

But of greater importance are the nerves and muscles of the pharynx.

There is scarcely any organ in the body connected with so numerous and various nerves as the pharynx.

The sensory nerves, which control on the one hand ordinary sensibility and sensation, and on the other taste and reflex muscular contraction as well as secretion, spring for the most part from the trigeminus. The lesser palatine nerves arise from the second branch and the spheno-palatine (Meckel's) ganglion, and extend from the anterior surface of the soft palate to the borders of the anterior pillars of the fauces. The gustatory parts of the pharynx, the lateral parts of the soft palate, and the anterior pillars of the fauces, receive their supply from the glossopharyngeal, while the secretory branches lie in the chorda tympani.

Motor innervation is, however, much more complicated.

The motor fibres are partly derived from the third division of the fifth nerve; the lesser superficial petrosal nerves, and the internal pterygoid branches, supply the tensor veli palati. The facial supplies the levator veli palati and azygos uvulæ by means of the superior palatine branch, and the pharyngo-palatine and glosso-palatine muscles through the inferior palatine branch.

The spinal accessory supplies in some persons completely, in others only partly, the superior constrictor of the pharynx; it alone sends fibres to the middle constrictor, and also to the levator veli, azygos uvulæ, and pharyngo-palatinus.

The centre for the first and voluntary part of the act of swallowing is, according to the experiments of Krause, in the gyrus prefrontalis of the parietal lobe of the brain.

Of all the muscles of the pharynx the constrictors are the most important on account of their function.

They are three in number, superior, middle, and inferior. By their combined action the pharynx is narrowed and the food propelled downwards.

The muscular bundle arising from the anterior pillar of the fauces, called the glosso-palatine muscle, forms, with the pharyngo-palatine muscle, which extends from the posterior pillar to the velum, a sphincter, which cuts off the mouth from the pharynx, and approximates the two pillars of the fauces. A very important function is that of the levator veli palati. Its action on the one hand is to open up the lumen of the Eustachian tube, and on the other to raise the soft palate upwards, and draw it against the posterior wall of the pharynx, and so to shut off the nasal part from the remainder of the pharynx.

To the tensor veli palati, according to Rüdinger, belongs only the small function of stretching the velum; but besides this, it at the same time helps in opening the tube.

The azygos uvulæ shortens the uvula in its long axis, and raises it, throwing the mucous membrane into diagonal folds.

Of special interest is the distribution of the glands between the muscular bundles on the oral surface of the velum palati. Rüdinger has shown that the individual glands are surrounded by oblique and transverse fibres, and that between the glands and the mucous membrane there is a special muscular stratum, consisting of fibres running obliquely; through it the muscles of the velum act, compressing the glands and causing them to secrete.

EXAMINATION OF THE PHARYNX.

On account of the peculiar conformation and relatively large dimensions of the pharynx, it is easily understood that ordinary inspection of it is not sufficient for its thorough examination. Nevertheless it constitutes the most frequent and most important method of examination. In the first place a suitable light is required, and as a rule daylight suffices. If, however, the person to be examined is confined to bed, and too far from the light, or turned away from the light, then out of consideration to the patient, who is unable to change his position, reflected daylight must be used; but if the weather be gloomy, or if it be at night, then reflected artificial light must be brought into requisition. The method of directly illuminating the pharynx, which is still so often practised, is not only insufficient but troublesome.

As few persons are able to keep their tongue down on the floor of the mouth by muscular power, it becomes necessary to hold it down by means of instruments for the purpose. In a case of need one may use the handle of a spoon, a paper folder, the fore finger, the handle of a laryngeal mirror, or a mouth spatula, of which there are a great number, some being straight, others angular, bow shaped, bayonet shaped, there being also those invented by Türck, Fränkel, and others. To whichever of them the preference be given, the rule is, that the broad surface be slowly and carefully placed upon the middle of the posterior part of the tongue, when, with increasing force, the tongue is pressed forwards and downwards on to the floor of the mouth. Attempts at retching, which nearly always occur, are, for the examination of the deeper and higher parts, decidedly advantageous. During these movements one can observe the soft palate, the tonsils, and the posterior wall of the pharynx from the lower half of the Eustachian prominence to the epiglottis, and very often the epiglottis as well. At first most people make a mistake, in that they do not place the spatula on the middle of the tongue, or do not exert sufficient pressure, the consequence being that the tongue, rising up with astonishing strength, throws off the instrument as a horse its rider.

Refractory children are best managed by first fastening their head and arms with a towel, then taking hold of the nose and quickly introducing the spatula between the teeth, which they open during inspiration. Other kinds of instruments, such as wedges, metal fingers, or gags, are unnecessary for examination and ordinary manipulations, but are of great use, especially in children, when tonsillotomy or longer operations have to be performed. Adults with false teeth, or those who are afraid that they will be infected by the instruments of the surgeon, frequently offer stubborn resistance to examination. With regard to the last, the beginner cannot be too careful in keeping his spatula, mirror, and brushes, absolutely clean.

In inspecting the throat one should notice quickly and at a glance the configuration, colour, and movement of the soft palate, the uvula and the pillars of the fauces, also the size and appearance of the tonsils, the condition of the posterior wall, and any secretion that may be present. One must never neglect to examine carefully the lateral walls of the pharynx, especially the part lying behind the posterior pillar of the fauces, and the folds of mucous membrane proceeding upwards and downwards from it.

For the inspection of the deeper parts of the pharynx a laryngoscopic examination is necessary, and for that of the naso-pharynx a special method is required.

Although it cannot be denied that pharyngoscopy or posterior rhinoscopy, on account of its great difficulty, has not received the same attention as laryngoscopy, nevertheless its importance cannot be too strongly urged.

The difficulties lie, on the one hand, in the anatomical structure of the naso-pharynx, and on the other in the irritability of the persons to be examined. One may pretty certainly decide by the inspection of the throat alone whether a person can be examined rhinoscopically, with ease or with difficulty, or not at all.

The greater the distance of the soft palate from the posterior wall of the pharynx, the easier and more complete is the examination; the less the distance, the more difficult it becomes. Although it is certain that with very sensitive persons it is impossible to accomplish all one's purpose, still, as one grows in experience the number of such cases diminishes every day.

The method of illuminating is the same as in laryngoscopy, *i.e.*, by means of the reflector. I use a small round flat mirror, $1-1\frac{1}{2}$ centimetres in diameter, fastened to a straight handle, almost at a right angle. I never make use of rhinoscopic instruments proper, *i.e.*, attached to a tongue depressor, or moveable mirrors and uvula holder.

EXAMINATION OF THE PHARYNX.

After regulating the light, the tongue is depressed with the left hand, so as fully to expose the soft palate and the isthmus of the fauces; in all cases the air bubbles between the arcades must be rubbed off with the handle of the mirror. Now the mirror, having been well warmed, is, with its reflecting surface turned upwards, carefully introduced with the right hand between the uvula and the pillar of the fauces, that is, into the right or left arcade, care being taken to avoid touching the back of the tongue or the pillars of the fauces; it is then pushed towards the posterior wall without touching it, while the patient stops breathing or breathes through his nose. The purpose of this manipulation is solely to keep the soft palate in the most relaxed position possible, and on this hinges the success of the examination. If in spite of this the soft palate approaches the posterior wall on account of efforts at retching, hence causing narrowing and eventually complete closure of the space for breathing to take place, then I allow the patient to try the effect of breathing deeply

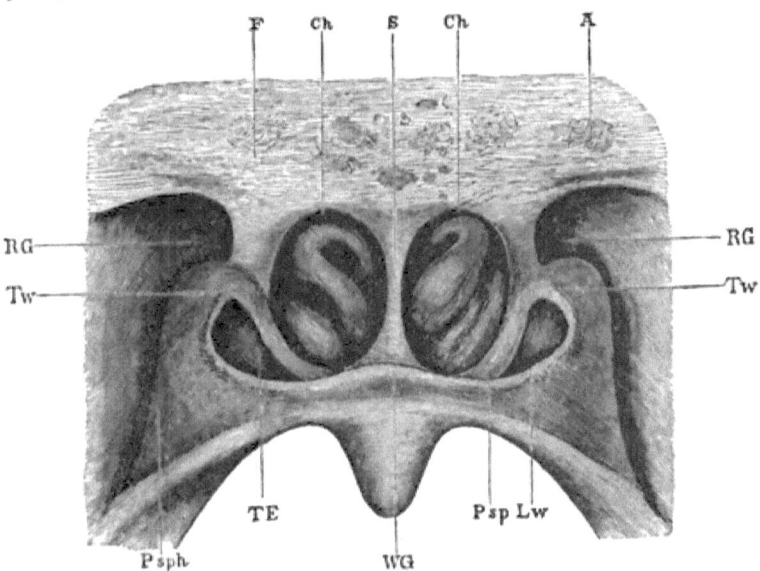

Fig. 1. THE RHINOSCOPIC IMAGE.

S. Nasal septum.
Ch. The nasal fossæ with the three turbinated bones and nasal meatus.
WG. Posterior surface of soft palate with uvula and place of junction of velum and hard palate; Levator in moderate contraction.
TE. Orifice of Eustachian tube.

Tw. Eustachian prominence.
Psp. Salpingo-palatine fold.
Psph. Salpingo-pharyngeal fold.
Lw. Levator cushion.
F. Fornix of anterior wall.
A. Adenoid tissue.
RG. Fossa of Rosenmüller.

through the nose; should this not now succeed in loosening the soft palate, I remove the mirror, and do not grudge the patient some rest, in order that the examination be undertaken for the second, third, or fourth time. In this way with adults and older children I nearly always succeed. The instruments which are recommended by various authors, such as palate hooks, uvula snares, uvula holders, the drawing of a silken thread or an india-rubber tube through the nose, which are all intended to enlarge the space to be inspected, I never use, because through them the irritability of the patient is only increased.

It is natural that by the application of so small a mirror the whole naso-pharynx cannot be examined at once, so that it becomes necessary to rotate the mirror in order to obtain a complete view.

The beginner should make his first trials on a lay figure, as, for example, Isenschmid's, and then go on first to the examination of healthy people with wide throats who are but slighly sensitive, and afterwards to that of patients.

It is somewhat difficult for the beginner to become thoroughly acquainted with the rhinoscopic image.

One always begins with the septum narium. Under ordinary circumstances it is seen as a pale red bony ridge prominently situated in the middle line. On its lateral borders the nasal fossæ are seen as two longish oval spaces: in these spaces one sees high up the smallest or superior turbinated bone, under it the larger or middle, and lowest down the largest or inferior, which last generally is somewhat covered by the contracted soft palate, so that only half or even less of it is seen; between these bones lie the superior, middle, and inferior nasal meatus.

Here may be mentioned the somewhat commonly occurring anomaly of the bulbous or even tumour-like thickening of the inferior posterior part of the septum. On rotating the mirror outwards, there is seen a well-marked roundish or oblong red swelling projecting from without inwards, beside which, at its inferior middle part, there is a pale sometimes unusually anæmic funnel-shaped opening; the former is the Eustachian prominence, the latter the entrance to the tube. The latter is bounded by two folds of mucous membrane, of which the anterior is called the plica salpingo-palatina, the posterior the salpingo-pharyngea; between them rises the prominence made by the levator. The part lying in the shade outwards from the Eustachian prominence is the fossa of Rosenmüller. The examination of the roof of the throat is the most difficult part of the whole; to do so it is necessary to hold the reflecting surface more horizontal, and raise the handle of the mirror well up. Under normal circumstances, the roof of the pharynx is seen as a slightly reddened uneven surface inclining backwards, or, from the presence of a well marked adenoid cushion, ragged and undulating. It is very easy to examine the posterior surface of the soft palate, which indeed only too often adds to the difficulty of getting a complete view of the parts described. It is seen in the mirror as a diagonal straight or undulating convex line running across the field of vision. It may be observed that the rhinoscopic image in very many persons is only partly or perhaps half visible, and that the appearance of the different parts, especially the turbinated bones and the Eustachian prominences, is subject to numerous deviations.

The possibility of inspecting a part of the naso-pharynx from in front through the nose cannot be denied; in patients affected with atrophy of the turbinated bones, one is generally able, without any further preparation, such as dilatation of the nostrils, clearly to recognise the Eustachian prominences with the openings of the tubes, the fossæ of Rosenmüller, and the posterior wall of the pharynx with a part of the adenoid tissue; and still more distinct is the view if the posterior part of the septum has also been destroyed. It is easy to understand that, in complete absence of the soft palate, or in large perforations and in clefts of the same, the clearest view may be had. Under normal circumstances it is possible to see through the nose only a small elongated sometimes fissured strip of the posterior wall of the pharynx, and very often not even that. On account of the difficulty of posterior rhinoscopy, Zaufal has introduced into practice a new method of examining the naso-pharynx by means of specula.

These are thick metal tubes, from 10 to 12 centimetres long, with a funnel shaped opening, which are carefully pushed, guided by the eye, along the inferior nasal meatus into the naso-pharynx, without disturbing the posterior wall. For illumination, the forehead reflector is used, with artificial light. Very seldom has any method been so variously criticised as this one. By some it is extolled to excess, by others completely thrown aside, but a middle course is the right one. It cannot be denied that the method may give valuable information about the Eustachian prominences and the openings of the tubes, as well as the folds of mucous membrane proceeding from them. But it is also true that the introduction of the speculum, even in quite straight noses, is, to say the least, very unpleasant, and even painful, and is often accompanied by bleeding, notwithstanding all that Habermann has done to defend this method, in opposition to Baginsky.

As a further method of examination, palpation with the finger, or with the probe, must be considered.

Although this method has in some degree been superseded by others, nevertheless, with many patients, and especially with children, it is unrivalled. And for completing and supplementing the examination, for searching for foreign bodies, for determining the situation, extent, and consistence of tumours, abscesses, etc., it is indispensable. It requires considerable practice to be able to give an exact account of what is felt, as the examination must be made as quickly as possible, yet at the same time carefully. Digital examination of the naso-pharynx, even when carried out with the greatest caution, gives rise, as a rule, to bleeding and great pain.

The person to be examined is placed on a chair, the surgeon steadies the head with his left hand, and then, while the patient inspires, passes the forefinger of his right hand as quickly as possible between the uvula and the pillar of the fauces, on to the posterior surface of the soft palate, which immediately contracts on, and grasps the finger. In a few seconds the contraction will cease; but if it should not, then the surgeon, in spite of it, must force his finger along the posterior wall toward the roof of the pharynx. In this way the Eustachian prominences and the inferior turbinated bones projecting somewhat into the pharyngeal cavity are felt, and the condition of the fornix ascertained. He then passes his finger along and down the other side, and withdraws it. It must specially be noticed if the space between the Eustachian prominences, as well as the roof of the pharynx, be smooth and even.

GENERAL THERAPEUTICS.

The remedies which are used in the treatment of diseases of the pharynx may be divided into two groups: those which are applied to the skin, and those which are applied to the mucous membrane.

Among those applied externally, the principal are fomentations or poultices, which may be classified as cool or cold, and stimulating or warm.

The cool or cold applications are regulated either by frequently changing compresses wrung out of perfectly fresh or cold water, or by putting an ice bag, made like a cravat, round the neck. The latter is a longish indiarubber bag, of the breadth of the hand, which is filled with small pieces of ice, or with cold water. Some have frequently doubted the power of this apparatus to lower the temperature, and have entirely discarded it. It is true that, owing to the extreme vascularity of the cervical region, the temperature of the deeper parts, more especially the mucous membrane, can be lowered only after prolonged application of cold; still, as the beautiful experiments of Winternitz show, the fact remains, that through the contraction of the larger and smaller arteries of the neck, the blood stream to the ramifications of the carotid, and to the component parts of the pharynx and larynx, may become lessened and the temperature diminished. Cold applications, and the use of ice inwardly, in the form of ice pills, ices, iced lemonade, etc., are indicated in all acute inflammatory and irritative processes of the walls of the pharynx and larynx, more particularly in phlegmon, croup, and diphtheria.

Much more popularly and widely used are warm or stimulating fomentations. They are used in the simplest manner; a handkerchief or towel is dipped in cold water, then well wrung out and put round the neck, with some impermeable material, such as gutta-percha tissue, oiled silk, etc., or simply a dry woollen cloth, laid over it. There can now be obtained specially prepared compresses, which require only to be dipped in water and bound round the neck. They need to be renewed only when the moisture has evaporated and the compress is almost dry. According to Winternitz, the question whether it is better to put an impermeable water-tight covering over the compresses, or simply a dry cloth, is answered in favour of the latter. Apart from the fact, that the dilatation of the blood vessels of the skin, caused by allowing the cloth to dry slowly, without an impermeable covering, appears to be a greater one, it is found that when watertight material is applied over the cloth, it interferes with perspiration, and often causes an unpleasant, uneasy feeling, or a troublesome skin eruption, and that after a certain time the hyperæmia disappears. Hot stimulating applications are obtained by dipping cloths in hot water or chamomile tea, or in the form of cataplasms or poultices. These applications produce, by dilatation of the vessels of the peripheral layers of the skin, a lowering of temperature in the inflamed parts of the mucous membrane, and also favour the transpiration of secretion through it. They are indicated in all cases in which it is desired to promote the resorption of inflammatory products, and also in the later stages of acute inflammation. Hot fomentations, poultices, etc., are used chiefly for maturing pus formation in cases of phlegmon of the pharynx, or of the peritonsillar or retro-pharyngeal connective tissue.

Applications of hot oil and fat, which are in so great favour with both educated and uneducated people, and also of the rind of bacon, are very unpleasant, and cannot in the least be compared with water compresses.

Inunction of salves, especially those containing iodine, are, on account of their property of absorption, not to be despised. Rubbing the skin of the neck with irritating materials, blistering salves, and the use of blisters, mustard poultices, etc., are but seldom ordered now-a-days. Local anæsthesia of the mucous membrane, by means of the ether spray, or by the injection of morphia near the superior laryngeal nerve, I have never been able to produce.

Of undoubted value is the cutaneous, or the endo-pharyngeal application of electricity, by means of the constant as well as the induced current.

On the other hand, massage is of very doubtful value. Valuable as it is in affections of the joints, severe injuries, etc., it accomplishes but little in the treatment of affections of the pharynx, although some enthusiastic supporters of it believe that they will be able to cure angina and enlarged tonsils by means of it.

Bleeding is now but seldom resorted to. It has been shown for a certainty that venesection in diseases of the pharynx and larynx is useless. For local blood-letting the use of leeches must first be considered. They are usually placed in the angle between the under jaw and the mastoid process. Opinions as to their value differ very much. According to Morell Mackenzie, from 3 to 6 at least must be applied on each side, if any benefit is to follow.

Scarification of the mucous membrane can only be done on the tonsils or posterior wall, on account of the formation of the pharynx; it cannot be denied, however, that through it great alleviation may be obtained.

Among the remedies which are applied to the mucous membrane of the throat, gargles are still considered the most important. When one sees every day in what diseases of the pharynx and larynx gargles are ordered, and used for months by the patient without any benefit, one cannot but be astonished at how little in general is known exactly of their therapeutics. It should at least be known what parts of the throat are rinsed out by them By numerous experiments, which I made on myself with coloured substances, I was able to determine with the greatest possible exactness, that the anterior surface of the soft palate, the tonsils, and a small portion of the posterior wall of the naso-pharynx, varying in extent according to the construction of the pharynx, are the parts which really come in contact with the gargle.

Hence it would appear that gargles are only indicated in affections of the organs forming the isthmus of the fauces.

The temperature of the gargle is not a matter of indifference. In general it may be held that in all acute, and especially in phlegmonous processes, the fluid should be used cold or mixed with ice; as the symptoms diminish, or if the cold becomes unpleasant, the temperature of the gargle should be raised, until it is lukewarm or very warm, and this is particularly the case in maturing abscesses or in encouraging suppuration.

Another very general method of treatment is the application of atomised fluids by means of inhalation apparatus or sprays. While some consider this method indispensable, others reject it as utterly useless. Although I myself hold that, by means of inhalations alone, one is unable to cure chronic affections of the throat, still I constantly use them as palliative measures for moistening and cleansing the mucous membrane, as well as for disinfecting it.

In chronic affections I use the cold or ball spray in preference, because, on the one hand, the fluid to be inhaled can be so easily regulated to any temperature, and on the other, because the mass of fluid coming in contact with the mucous membrane is greater than can be produced by steam inhalers. I recommend very specially those atomisers which have a long mouth tube with a nozzle, which can be turned both upwards and downwards, so that both the naso-pharynx and the larynx are sure to be reached by the fluid. The simplest method of cleansing the naso-pharynx is by snuffing up through the nose a fluid containing common salt or soda, the head being thrown well back. The other methods will be spoken of under diseases of the nose.

Among the most effective methods are painting and cauterising the mucous membrane with fluid or solid substances.

For the former purpose, a brush on a straight wooden or bent metal handle, is used, made of soft badger's or marten's hairs. The result depends not only on the choice of substances, but also on the energetic and skilful use of the brush, which generally is too timidly handled, and sometimes is too small. Of course the brush for the laryngeal part of the pharynx must be larger, and that for the naso-pharynx have a shorter bend. For the purpose of cauterising, it is very useful to have a small piece of sponge or wadding held in a forceps, or attached to a wooden handle.

When cauterising with solid nitrate of silver, one must make sure that it is properly fastened in. Beginners ought not to use the ordinary caustic holder, because of the danger of the caustic breaking off; it is better to use an iron wire, to which each time it is used a small piece of caustic is made to adhere by dipping it into nitrate of silver melted in a platinum dish. Hering has lately recommended the use of chromic acid.

Remedies can further be applied in the form of a powder. To apply these a powder blower or insufflator is used. This has at one end either a ball, or a piece of india-rubber tubing half a metre long provided with a mouth-piece, and has near the middle an aperture, shut by a sliding tube, into which the suitable medicament is put. Bresgen's insufflator is a specially good one; it is provided with a valve, and can be taken to pieces.

On account of the small choice of powerful non-deliquescing caustics, the introduction of the galvanic cautery into the therapeutics of diseases of the pharynx, larynx, and nose, must be regarded as a most praiseworthy advance. The indications for it will be found more fully under the different diseases in which it is used; here there is only room for a few remarks about the instruments.

Without in the least undervaluing the services of older scientists, especially Bruns, Hedinger, and Voltolini, the author of this book believes that his handle for galvano-caustic operations on the nose, pharynx, and larynx, invented in 1877, has gone far in supplying a decided want. The author can, at least, flatter himself that his instruments of all sorts have been introduced into England by Morell Mackenzie, Semon, and others.

Figure 2 shows the author's universal handle, into which can be fastened not only all the solid cauteries, but also all the tubes for

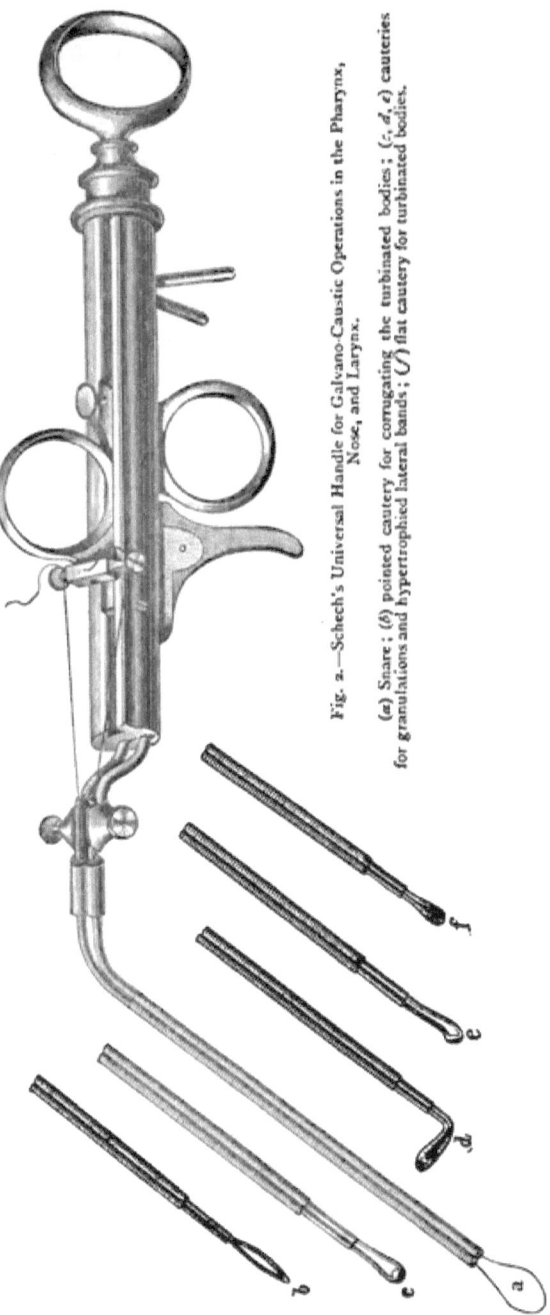

Fig. 2.—Schech's Universal Handle for Galvano-Caustic Operations in the Pharynx, Nose, and Larynx.

(*a*) Snare; (*b*) pointed cautery for corrugating the turbinated bodies; (*c, d, e*) cauteries for granulations and hypertrophied lateral bands; (*f*) flat cautery for turbinated bodies.

Fig. 2.

the snares. The revolving ivory ring behind is for the reception of the thumb, the fixed ring below, also of ivory, is for the third, the trigger-like apparatus in front of it for the fourth, and the moveable ring above, with the slides and transverse pieces, for the index finger. The conducting wires appear on the under surface; by them the weight is to a certain extent equalised, so that perfect control of the instrument is obtained. Should the handle be used for the galvano-caustic knife, the slide is fixed by the screw behind the ring, while the ring forms a support for the index finger.

In figure 2 there are also represented several of the instruments most commonly used in operating on the throat and nose: *a*, represents the snare for the nose; *b*, a pointed cautery for corrugating or touching the turbinated bones; *c*, *d*, *e*, cauteries for destroying granulations and lateral bands; *f*, a flat cautery for the turbinated bodies. A case of these cauteries may be had from Eugen Albrecht, university mechanic at Tubingen, for 80 marks (£4). It contains fifteen in all, excellently finished, and of all possible forms and curves, for the throat, nose, and larynx. The second handle, to which the solid cauteries are fixed, is very useful in difficult operations on the larynx, when it is necessary to hold the instrument like a pen.

Of the numerous galvano-caustic batteries, I particularly recommend Brun's zinc and carbon battery, which is charged with chromic acid. It is sold by Albrecht for 90 marks (£4, 10s.). I can also recommend the battery made by Voltolini.

The deficiency of all batteries, their small degree of constancy, the trouble involved in filling, and the way they so soon get out of order, make it very desirable that cautery heat be produced in some other way. With the extraordinary advance electricity has made in the last few years, we hope that the difficulties are not insurmountable. So successful are the results published by Bayer, of Brussels, with Faure's accumulators, that they only require to be more experimented with to become introduced into general use.

MALFORMATIONS AND DEFORMITIES, CONGENITAL AND ACQUIRED, AND OTHER ANOMALIES.

The congenital defects of most common occurrence are, clefts or fissures. They extend from the incisor teeth to the hard or soft palate, and are known by the name of "*Wolfsrachen*" or *cleft palate*. Lateral clefts of the palate are also seen, which affect either the bone only or the soft parts.

Many anomalies of form and functional disturbances, too, may be caused by abnormal processes in the bony walls of the pharynx.

Beginning with the base of the skull, we first meet with abnormal length of the styloid process. Such cases are described from an anatomical point of view by W. Gruber, and from a clinical by Lücke and Weinlechner. This abnormal length of the styloid process, and the ossification of the styloid ligament, by reaching as far as the posterior surface of the tonsils, or by being bent to one side, may give rise to abnormal sensations and continual impediment to swallowing. By breaking away the offending osseous process, Weinlechner was able in one case to remove these inconveniences.

Difficulty in swallowing is not infrequently caused by exostoses of the cervical vertebræ (Scheff), also by lordosis (Stoerk), or by defective formation of one of the cervical vertebræ, with disease of the spinal column (Heyman). These difficulties may, in some cases, be removed by wearing a suitable cravat.

Turning now to abnormalities of the soft parts of the pharynx, we meet with congenital or acquired defects of the whole of the soft palate, congenital perforation of the anterior faucial pillar, and absence of the uvula. Of the acquired anomalies, the perforations of the soft palate deserve mention, and will be spoken of later on.

The uvula may be congenitally very small or unusually large. It may be cleft either alone or along with the soft palate, and any degree may occur, from the ordinary indentation to the pronounced cleft palate. *Uvula bifida* must not, however, be confounded with this; in it the ends diverge from one another, like the tail of a lyre, or overlap. In the latter case they must be separated with a probe, in order to ascertain how high the cleft reaches toward the base.

Complete clefts of the uvula may occasion nasal timbre of the voice, from the pharyngo-nasal isthmus being too widely open, a symptom which occurs to a high degree in cases of the previously mentioned perforations and defects. The larger the opening the greater the difficulties of deglutition; the food generally being regurgitated. The uvula is very often split, distorted, or drawn in various directions, in consequence of previous ulcerations. Further anomalies are seen in the absence of the tonsils or their rudimentary development; comparatively often the deposition of a part of the tonsils on the posterior pillar of the fauces occurs, which thereby appears uneven, and covered with yellowish red masses.

Of acquired defects, narrowing of the pharynx, in consequence of cicatricial contractions, is one of the most common.

As etiological factors, the ulcers caused by syphilis, lupus and diphtheria, also injuries, scaldings, and cauterisation of the mucous membrane, are to be taken into account.

The naso-pharynx is in rare cases separated from the nasal fossæ on one or both sides by a lamella of bone (osseous closure of the nasal fossæ), and division of the naso-pharynx has been observed from the septum narium being continued back as far as the vertebral column. Schrötter describes several cases of partial and total closure of the nasal fossæ by membranes, the results of syphilis, and blennorrhœa (first described by Störck). Complete closure of the Eustachian tubes often occurs; and sometimes, in consequence of violent nasal expiration, the fossæ of Rosenmüller are symmetrically dilated (Pertik).

We are interested, however, in the adhesion of the soft palate to the posterior wall of the pharynx.

Partial adhesion on one or both sides is the form that most frequently is met with. The adhesion begins as a rule from the inferior part of the posterior pillar of the fauces, and extends from there to the middle of the soft palate, and also to the uvula. The simplest form is that which takes place between the posterior surface of the posterior pillar and the pharyngeal wall. It is easily understood that by these adhesions the pharynx comes to assume most manifold and curious forms. Somtimes along with the adhesion, and more or less distinct horizontal position of the pillars of the fauces, there may occur displacements on account of cicatrices of the posterior wall. Very often the uvula is intact, but behind it remains an aperture which is scarcely big enough to admit a pencil.

Disturbances from partial adhesion limit themselves mostly to the speech, which sounds toneless and nasal. Obstructed respiration through the nose also occurs, and more rarely difficulty of deglutition.

Complete adhesion of the soft palate with total closure of the naso-pharynx is of the rarest occurrence, as the collected writings of Kuhn show. Through complete closure, smell and taste are lost; there is also a certain idiotic expression of countenance, on account of the mouth being kept open; the mouth is dry, the tongue furred, and sleep restless. Speech is not only nasal, but also often nearly incomprehensible. In many cases the Eustachian tubes are at the same time obstructed.

On turning to the anomalies of the posterior wall of the pharynx, we find mostly deep white cicatrices and raised ridges as results of syphilis, proceeding upwards and downwards from the naso-pharynx. I once saw the posterior wall of the pharynx, in consequence of syphilis, divided into two halves by a central hypertrophied cicatrix, which had the form of a crest or ridge : in another case, the uvula was separated from the velum, and was adherent to the posterior wall : in many cases it becomes fixed to the posterior surface of the soft palate.

In the lower divisions of the pharynx, contractions or strictures are of much rarer occurrence.

As I have already stated in an earlier work, the appearances exhibited by these deep lying contractions are very varied, and can only be understood after repeated examination, especially when proliferations and constrictions are present on the tongue or on the edges of the cicatrix. The simple forms are those in which the pharynx to a greater or less extent is bridged over by a membrane reaching from the posterior wall or the lateral parts to the base of the tongue. A small aperture of the size of a pin head, either in the centre or at the periphery, may remain, through which can be seen, as in the case described by Langreuter, either none at all, or only a part of the normal or altered larynx. Sometimes a part of the epiglottis, compressed till like a catheter in shape, protrudes from the opening, sometimes the edges of the epiglottis become adherent to those of the opening, and can scarcely be recognised at all, from thickening and distortion.

The cases in which cicatricial bands, resembling bridges or trellis work, stretch across the pharyngeal cavity, or like "side scenes" protrude into its lumen, present very complicated appearances. The disturbances brought about by these different constrictions mostly affect deglutition. Apart from the impossibility of allowing large pieces of food to pass through, very often the patient is not even able to swallow small ones, or if he does it is only at the expense of much time and trouble; because, on the one hand, the greater part of the piece remains between the cicatricial band and the pouches formed by it, and on the other because the mucous membrane being changed into cicatricial tissue, numerous glands for the moistening of the pharynx are atrophied, and the muscular fibres under the mucous membrane have become incapable of contracting. In very extreme cases the patient can only partake of fluids; still, with practice and ingenuity, as I have repeatedly observed, the impediment to swallowing may in time be partly overcome. Difficulty of respiration must of course occur in all those cases in which the communication with the larynx is considerably obstructed. Asphyxia may occur from too large a piece of food sticking in the aperture of communication.

The prognosis of pharyngeal stenosis caused by cicatricial contraction, depends upon the degree and the situation of the lesion, but it has become much more favourable than formerly, on account of the perfection of laryngoscopy. Nevertheless a proportion of the cases must always remain uncured, or only partially improved.

Treatment.—The treatment of stricture of the pharynx should, in the first place, when possible, be prophylactic; by careful general and local treatment of the ulcers, and by early application of tampons, it will be possible to prevent adhesion of opposing epithelial surfaces, or to break down recent adhesions. When the stenosis is of longer standing, other methods must be attempted.

In cases of closure of the Eustachian openings, the forcible blowing in of air, as well as the introduction of bougies, and the incising the anterior wall of the tube, have been recommended, but unfortunately all these methods of procedure are untrustworthy. When there is complete obliteration of the tubal apertures, the production of a permanent opening in the drumhead, with extirpation of the hammer, and application of the air douche, are recommended to remove the deafness.

Closure of the nasal fossa, when osseous, can be overcome through the nose by means of a trocar, and prolongation of the septum posteriorly by breaking it off with bone forceps.

Membranous adhesions of the nasal fossæ are best removed by the galvanic cautery introduced through the nose or from the pharynx, and then kept open by passing in metal bougies. For this purpose Schrötter tied the bougies in front with a string, so that a constant pressure may be exercised on the lateral wall of the artificially constructed aperture, by their posterior ends diverging from one another.

Partial adhesions of the soft palate require no treatment, since as a rule they give rise to no symptoms.

It is quite otherwise, however, with total adhesions. In these cases the soft palate must be separated from the posterior wall, the edges cauterised, and recurrence of the adhesion prevented by introducing bougies. Thick caoutchouc drainage tubes drawn through the nose, or as Kuhn recommends a hard caoutchouc or silver plate convex anteriorly, occupying the whole breadth of the pharynx, which by a spring apparatus can be fastened to two opposite upper back teeth, are the best instruments for this purpose. Lieblinski arches the soft palate forwards by means of a catheter introduced through the nose, notches it, and passes through the gaping slit suitable bougies or leaden probes, which are allowed to remain in a longer time each day.

The excision of the soft palate is indicated, according to Ried, in extensive superficial adhesions, or in cases where separation of the soft palate would be too difficult: complete adhesion of the velum with perforation is best let alone.

The results as regards restoration of nasal respiration are generally favourable, still there is a decided tendency to repeated narrowing, so that a second operation is often rendered necessary. On the other hand, the results as regards improvement and restoration of speech are generally very unsatisfactory. The cause lies principally in the difficulty in restoring to the soft palate its normal length and mobility particularly when the muscles are atrophied and the mucous membrane has become unyielding from cicatrices of the connective tissue.

The treatment of membranous adhesions of the deeper parts of the pharynx depends upon the degree and form of the stenosis. The simplest is the separation and excision of the membranes from the existing aperture, with a probe-pointed knife or a pair of scissors, conducting the operation with the laryngeal mirror. To remove larger protuberances and vegetations, the electric snare is most suitable. The after treatment consists in careful cleansing of the edges of the wound, and the introduction of Schrötter's dilating tubes for gradually lengthening periods. Relapses, however, may occur.

With regard to the treatment of cleft palate and the other congenital anomalies, text-books on surgery must be referred to.

ACUTE CATARRHAL PHARYNGITIS.

Synonym.—Pharyngitis catarrhalis acuta.

The pharynx, as is the case with the mouth, is more often attacked by catarrhal inflammation than by any other affection. By catarrhal pharyngitis (angina catarrhalis), is understood a superficial inflammatory process, characterised by redness, velvet-like swelling of the mucous membrane, and increased secretion. It often increases in individual parts of the mucous membrane, especially in the glands, and goes on to suppuration, and then represents a combination with phlegmon.

The exciting causes of acute catarrh of the pharynx are very numerous. Chills, draughts, getting the skin of the neck or of the throat wet, or any other part of the body, and especially the feet in some persons, are recognised by all as amongst the most common of exciting causes. Hence one meets with acute pharyngeal catarrh associated with acute rhinitis, stomatitis, laryngitis, and bronchitis; also with muscular and articular rheumatism, and especially wry neck and lumbago. To the primary or idiopathic catarrhs belong those which are caused by partaking of too highly seasoned or too hot foods and drinks, or by breathing hot air.

As a secondary or symptomatic symptom is the acute catarrh of the pharynx, which usually follows infectious diseases, especially measles, scarlatina, roseola, smallpox, typhoid, and syphilis; it also occurs after taking certain medicines and poisons, especially preparations of belladonna (atropin), iodide of potash, mercury, antimony, and some acids and alkalies (*toxic angina*). *Angina arthritica*, described by Morell Mackenzie, as occurring in the course of an attack of gout, seems to be almost unknown on the continent.

The whole of the mucous membrane of the pharynx, or only a part of it, may be the seat of the inflammatory process.

Should it be localised to the naso-pharynx, objective examination, after any superabundant mucus has been removed, shows more or less redness and swelling of the mucous membrane, especially of the glandular cushions on the posterior and superior walls. The projecting edges also of the Eustachian tubes, and the folds of mucous membrane extending from them, share in the inflammatory swelling, which spreads to the posterior surface of the soft palate and to the uvula; the mucous covering of the latter appearing rough on account of the swelling of the follicles. Sometimes the follicles suppurate in large numbers, and impart to the mucous membrane a speckled appearance.

The pharyngo-oral cavity, and the parts forming the isthmus of the fauces, are particularly often affected. If the mucous membrane covering the vertebral column be affected, there are often on it alone, light or deep red streaks, but the posterior wall generally assumes an intensely red surface. At the beginning of acute exanthemata, the mucous membrane appears mottled and spotted, while later on the spots run together; in like manner, the mucous membrane at the commencement of the disease appears dry and covered with scanty secretion, while at a later period it becomes invested with a thick, tough, vitreous, and mucopurulent secretion, and is also rough on account of the swelling of the glands, while here and there is seen a follicle larger than usual, with a yellow purulent top, or a round cup-shaped follicular ulcer.

The interarytenoid region, the aryepiglottic ligaments, and the vocal cords, often exhibit the same changes.

The soft palate, too, generally shares in the inflammation. It appears thickened, more or less reddened, and distinctly separated from the paler hard palate. On the reddened region are numerous ramifications of vessels, dilated veins, and even little hæmorrhages. The uvula is enlarged, both vertically and across; the inferior portion of it is œdematous, its tip inclined to one side, or bent like a horn, and its mucous covering rough and granular, through swelling of the glands. Very often the affected side of the palate is more depressed than the sound one, and during phonation is either quite immoveable, or is drawn over to the affected side.

The tonsils are often affected alone, along with the posterior wall, or with the palate.

The simplest form of catarrh of the tonsils, is characterised by uniform reddening of the mucous membrane, and more or less considerable swelling and secretion of mucus. The condition is quite different when the inflammation penetrates into the lacunæ. On the uniformly reddened and swollen mucous membrane, there appear whitish or yellow spots, which are oval, three-cornered, or distinctly stellate; sometimes they assume a linear form. These correspond exactly to the number, size, form, and distribution of the openings of the lacunæ. They are composed of swollen and shed epithelium, muco-pus, and fungi, and can be removed without much difficulty.

Another form, or rather a combination, is *angina follicularis*, with phlegmonous inflammation.

On one, but generally on both tonsils, which are reddened and enlarged, there are seen white or yellow more or less elevated spots and specks, which are either collected into groups or are equally spread over the surface; in some parts they run together. These spots are parts of the glandular parenchyma, and represent suppurating follicles; the surface of the tonsil being often uniformly covered with such spots. As the disease continues they are thrown off, and there appear, according to the depth to which the process has penetrated, superficial or deep cup-like ulcers, which quickly heal and leave behind retiform cicatrices, or saucer-like depressions.

Of the subjective symptoms, the most constant is fever, which appears at the beginning. Sometimes slight shiverings, general malaise, and headache; sometimes distinct ague, raised temperature to 104° F. or more, and a pulse of 120, or in children, 140; vomiting, restlessness, general depression, restless sleep, loss of appetite, increased thirst, set in along with pain on swallowing, which is generally very severe, especially when it attacks the posterior wall, but also when there is suppuration of the follicles. In acute post-nasal catarrh there is generally no pain; the parts are sensitive while speaking and swallowing, and breathing through the nose is interfered with; very often the ear is affected with tinnitus and deafness, along with noises in consequence of closure of the tubes, or of middle ear inflammation, with perforation of the drum head.

Dryness of the throat is constantly complained of; that, however, soon gives place to increased secretion of mucus, which causes repeated hawking and swallowing. When the larynx is affected, there is hoarseness and tendency to cough. The condition of the voice is very characteristic, speech is toneless, with nasal timbre, as if the patient had a foreign body in his throat. Pain on speaking, as well as regurgitation of food, is to a large extent due to the inability of the soft palate to perform its function. Patients often complain of pain outside the neck, which is caused either by co-existent rheumatism of the muscles of the neck, or by sympathetic swelling of the lymphatic glands. Cold in the head, with frontal headache, pain in the joints and loins, and conjunctivitis, often occur at the same time.

Acute pharyngeal catarrh generally disappears in a few days, but may, from unfavourable conditions, last a longer time—ten to fourteen days. One often meets with articular rheumatism setting in after the disappearance of acute catarrhal, but especially of follicular angina, and hence it is supposed that the former is dependent upon the latter. Some authors consider that the joint affection arises from the throat in a reflex manner, from irritation of the vasomotor nerves; others again trace both affections back to infection; but it is most likely that the injurious poison first develops in the throat, and then continues its action in the joints.

The diagnosis is easy to arrive at, from the objective condition. From the confusion raised by giving to every catarrh of the pharynx, but especially follicular ulceration of the tonsils, the name diphtheria, the differential diagnosis is of considerable importance. Since the subjective symptoms may entirely correspond with those of diphtheria, the objective condition must guide the decision. When purulent spots are uniformly distributed over the surface of the tonsil, it is possible to make a mistake only from gross ignorance, or intentionally; it is certainly difficult, sometimes even impossible, at the first examination, to distinguish whether or not the adherent elevated white spots have been formed by the coalition of several neighbouring follicles. The following diagnostic points, viz., the yellower or more greenish yellow colour of the spots, the slight tendency to their further spread, the limitation to the tonsils, the slight swelling of the lymphatic glands, the presence of unbroken prominent pustules, and the early appearance of funnel-shaped losses of substance on the site of the deposit all point with certainty against the affection being diphtheria.

Treatment.—The treatment must, in the first place, be prophylactic, on account of the extraordinary predisposition of many people to acute pharyngeal catarrh.

For sensible hygiene the foundation should be laid in early youth; daily experience shows how much this is neglected; and we see also that these very children who are most carefully protected from cold draughts, and are provided with as many clothes as possible to protect them from the weather, are specially often attacked with angina. This delicate upbringing and avoidance of draughts is continued systematically on to manhood, and is increased by the over-heating of rooms by servants and wise friends, until complete relaxation of the skin, and constant tendency to perspire, occur in a most critical manner. It may be granted, however, that weather-proof and hardy natures do not always defy changes and variations of temperature with impunity, although it is a fact that the stronger and more hardy the outer skin is, the more rarely are the mucous membranes attacked by inflammation. There can be no doubt that we can effect a change in circulation and secretion, and a strengthening of the cutis, by influencing the vasomotor nerves of the skin. To many who are predisposed to catarrh of the mucous membranes, a change of clothing is often sufficient to bring about such an effect. One does not at all require to be an absolute believer in the woollen régime which Jäger thinks can alone make one happy, in order to perceive that through the wearing of a flannel shirt, evaporation and cooling of the skin is really rendered slower or modified. In like manner it may be believed that, by lowering abnormally high temperature of rooms, a common cause of catching cold is removed. The skin may be directly strengthened by means of the different hydropathic methods, especially by daily washing with cold water, rubbing, swathing and douching, and in summer by river, plunge, and other baths. But the best treatment undoubtedly is living at the sea side and bathing daily, or in a well conducted hydropathic establishment.

Along with hardening of the skin, hardening of the mucous membranes may be conveniently combined by means of methodical gargling or syringing, for a month at a time, with fresh cold water, to which some astringent may be added, such as alum. It must be pointed out to the impatient, and to those who are always asking for some new remedy for their recurring catarrh, that relaxation of the skin, which has gone on for years, cannot be removed in a few weeks or even months.

To prevent the outbreak of a catarrhal and phlegmonous angina, or to check it, every kind of remedy has been proposed. In England, Morell Mackenzie recommends pastilles containing 3 grains of the resin of guaiac, to be administered on the first appearance of the symptoms, every two hours. They have proved to be very beneficial, and are very much used. Of less benefit is the administration of from 3 to 6 drops of tincture of aconite every three hours. When muscular or articular rheumatism accompanies the angina, salicylate of soda, in gramme (grs. 15) doses every four hours, quickly gives relief. Concato and Bufalini recommend, at the beginning of the angina, spraying with ether, by means of Richardson's apparatus, every two hours for three or four minutes. Vapour baths, and other means of producing perspiration, are also much thought of methods of checking catarrh. Although it cannot be denied that in some cases incipient angina or coryza may be prevented by excessive abstraction by the skin, still experience shows, that one generally arrives *post festum*.

When the disease has already broken out, attention must be paid to the existing fever. According to B. Fränkel, several doses of quinine, up to 1 gramme (grs. 15), given in the evening, will considerably shorten the course of the malady. By the application of ice to the neck, sucking ice pills, taking ice drinks, and by the administration of an aperient, which is far too often forgotten, one generally succeeds in moderating the symptoms and cutting short the disease. When there is pain on swallowing, pulpy or fluid food at a suitable temperature must be given. Since generally in from two to three days a certain intolerance for the application of cold, both externally and internally, begins, one must have recourse to warm and stimulating poultices, warm gargles, and warm drinks, especially alkaline waters, in order to accelerate the resorption of the inflammatory products, and make allowance for the subjective sensations of the patient.

As gargles and inhalations, and for rinsing out the naso-pharynx, the most suitable remedies are diluted lukewarm milk, Ems water, mucilaginous decoctions, solutions of glycerine 1 to 10, 1 to 2 per cent. solutions of bicarbonate of soda, chloride of sodium, or biborate of soda. When the secretion is copious, and the isthmus of the fauces is affected, astringent gargles of alum and tannin must be used; and in follicular ulceration, chlorate of potassium (grs. 20 ad. ʒ i).

In the latter cases, 4 per cent. solutions of boracic acid, and 1 to 3 per cent. solutions of carbolic acid, are also indicated, though what the action of these remedies is, can no more be explained than can the action of permanganate of potash in simple catarrhal angina.

In inflammation of the deeper parts of the pharyngo-oral and pharyngo-laryngeal cavities, inhalation of warm atomised remedies is preferable, although, however, only a small quantity of them can reach the diseased parts. If recovery be slow, then painting with glycerine of tannic acid, 1 to 10, or with nitrate of silver (grs. 50 ad. ʒ i) or with Mandl's solution of iodine (to be mentioned later on), will hasten the resorption.

The greatest care is required when the ear participates in the disease. On the first appearance of symptoms, the air douche must be used, and in a more serious attack, catheterism is necessary, which, however, is very painful when the disease is acute; as a rule, Politzer's method is sufficient, especially with children. Otitis media, as it develops, must be met with the usual remedies: leeches, application of ice, and eventually, if required, paracentesis of the drumhead.

Besides these usual forms of acute angina, there are various modifications which occur seldomer. Thus, for example, E. Wagner mentions slightly elevated white, or whitish yellow spots occurring, either circumscribed or diffusely, on the mucous membrane of the palate and tonsils, caused by suppuration of the epithelium. The outer layers of epithelium are generally awanting; the next are blistered and œdematous, and have pus and blood corpuscles scattered here and there throughout them. Sometimes the upper surface appears rough, uneven, and greyish, the epithelial layers are desquamating, and between them are seen little nodes of fat. This form has been designated *cachectic angina,—angina pultacea*. It appears only in states of debility of acute or chronic nature, *e. g.*, typhoid fever, phthisis, and senile marasmus. Whether it is identical with the disease described by Morell Mackenzie as *angina ulcerosa*, which occurs in anatomists, medical men, and sick nurses, in consequence of some slight septic infection, I do not venture to decide. There also may occur, following acute pharyngitis, smaller or larger ecchymoses, which form has been called *hæmorrhagic angina*, and which is described in the chapter on pharyngeal hæmorrhages. Inflammations of the pharynx following scurvy (*angina scorbutica*) and idiopathic thrush are also seen. On the reddened mucous membrane of the soft palate, uvula, and tonsils, elevated yellow spots appear after a few days, quickly developing into deep ulcers, which, if the patients be left to themselves, may give rise to extensive destruction and gangrenous desquamation. The odour from the mouth is fetid, and salivation copious; the subjective symptoms are the same as in acute pharyngitis and stomatitis. The treatment is similar to that of the corresponding diseases of the mouth. Chlorate of potash here also proves itself to be the best local remedy.

CHRONIC PHARYNGITIS.

Synonym.—Pharyngitis Chronica.

The causes of chronic pharyngitis may be divided into predisposing and directly exciting.

There is no doubt whatever that weakly, scrofulous, and tuberculous persons, also those suffering from heart disease, or convalescing from severe and especially from acute infectious diseases, likewise, according to Rühle, those suffering from hæmorrhoids, and persons affected with syphilis, are unusually predisposed to pharyngeal catarrh. By far the commonest of the directly exciting causes is the occurrence of attacks of angina, which either frequently recur, or rapidly follow one another. When one considers how seldom complete recovery from an attack of angina is waited for, one cannot be surprised, if after each attack some unabsorbed material remains behind, and acute catarrh resolves itself into chronic.

Chronic pharyngeal catarrh almost always accompanies chronic cold in the head, adenoid vegetations, and naso-pharyngeal polypi. More frequently careless living, and particularly the consumption of alcoholic drinks, pungent, and highly seasoned food, and smoking, must perhaps be blamed; while the fact that snuffing, and the still more loathsome habit of chewing tobacco, have the same effect, requires no explanation.

Certain occupations and trades also give rise to chronic pharyngeal catarrh: persons who work in a dusty atmosphere, millers, stone hewers, and file cutters, workers in metal, also joiners, and female workers in cigar factories, are almost as often affected by catarrh of the pharynx as by inflammation of the larynx.

In many persons, over-straining the voice is the cause of the affection, as is seen in its relative frequency among teachers, singers, actors, preachers, officers, and attorneys. Bresgen, who will not admit over-straining of the voice alone as an exciting cause, explains the apparently excessive tendency to this disease in those whose occupation is speaking, by the circumstance, that in their profession they become disturbed much earlier than other people by the consequences of their illness; on the other hand, however, he allows that catarrh of the throat already existing may really become worse by continued over-straining of the voice.

I am of opinion, however, that over-straining of the voice may of itself give rise to chronic pharyngitis, and especially to hypertrophy of the lateral bands, but I believe that it far most frequently develops its influence when allied with one or several other affections. To the author, who in his sphere of labour has sufficient material to decide the question at his command, the small number of female teachers, actors, and singers that participate in the affection, and the large proportion of officers commanding in the open air in all weathers, and in the dusty riding school, and who fully understand how to live well, appears most remarkable. Among teachers, who, generally from pecuniary reasons, rarely exceed, I could account for the origin of the affection, partly by anomalies of constitution and partly by pronounced predisposition to taking cold.

In only a general way therefore can one agree with Rühle, who blames the cigars, the wine, and singing for the great prevalence of throat affections in Rhineland, and accept his etiology, with some modifications on account of the quality of alcohol, and the more or less artistic use of the vocal organs.

Chronic catarrh of the pharynx may either attack the whole of the throat, or may be localised in separate parts; the process, however, is not usually sharply defined, but extends more or less over the various parts. Roth maintains that the pathological process always attacks the whole extent of the mucous membrane of the pharynx, and that a part cannot be affected by itself: to this statement, however, I must, from a clinical standpoint, give a most decided contradiction.

Hypertrophy of the parts affected, must be looked upon as a characteristic of all kinds of chronic catarrh of the throat, diffuse as well as localised: it extends on the one hand to the follicles and follicular glands, and on the other to the whole mucous membrane and the submucous connective tissue (*Pharyngitis chronica hypertrophica*). Since many inflammations of the pharynx terminate in atrophy of the mucous membrane, there must also be differentiated a *pharyngitis chronica atrophicans sive sicca*.

Chronic catarrh of the naso-pharynx presents changes which differ essentially from those observed in other forms; on this account therefore we will consider it first.

Hypertrophy of the tissue elements forms the substratum of chronic naso-pharyngeal catarrh. The mucous membranes covering the Eustachian prominences, the fossæ of Rosenmüller, and the posterior surface of the soft palate, appear thickened, and more or less hyperæmic: the blood vessels are dilated, and sometimes the parts are of a greyish white colour, from proliferation of epithelium. The mucous glands and the follicles stand out more distinctly, the posterior ends of the turbinated bodies are hypertrophied, or resemble small bladders, or frog spawn. The mucous membrane of the naso-pharynx may be normal; it is, however, much oftener also hypertrophied and covered with granulations, while the tonsils are enlarged.

The adenoid tissue is attacked with special frequency and intensity. When hypertrophied, the result is known as "adenoid vegetations," or as "hypertrophy of the pharyngeal tonsil."

In well marked cases, the impossibility of seeing the nasal fossa is what strikes the examiner most. However much he turns the mirror, nothing can be seen of the characteristic septum narium, or of the turbinated bones. Instead of these he sees a reddish or greyish red mass, made up as it were of cones, berries, or stalactites, reaching from the roof of the pharynx to the junction of the soft and hard palates, and to the openings of the Eustachian tubes, sometimes even covering them and filling up the lumen of the naso-pharyngeal cavity. The mass is more or less covered with a thick, tough, greenish yellow, or bloody secretion, here and there forming scabs. On removing these one sees that the individual swellings, or crest-like prominences, are separated from one another by deep, generally longitudinal slits, and are more or less pedunculated, and moveable to a greater or less degree. Wilhelm Mayer, to whom the greatest merit is due for having first described this disease in detail, divides these growths into leaf-shaped and cone-shaped, and as sub-divisions he mentions the crest-shaped and plate-like. The Eustachian openings are either covered by a large tumour-like mass, or by several small excrescences like cocks' combs, or are partially seen.

In less severe cases, only the inferior sections of the nasal fossæ and septum can be seen, while the superior are covered over by the above described tumours. The Eustachian openings may be free, while single tumours may extend as far as the fossa of Rosenmüller, or into the immediate neighbourhood of the tubes.

Adenoid vegetations are of but slight consistence; in children they are very soft, sanguineous, and delicate, and may be torn or injured by the palpating finger. In adults the surface is generally equally soft, but the base is hard and firm; they are especially hard when situated on the cartilage of the Eustachian tube, or on the posterior surface of the soft palate.

As in size and number, so also in their seat and point of origin, they vary considerably. Sometimes they spring from the fornix, both from the anterior and posterior walls; in rare cases the fornix is quite unaffected, while the posterior wall, as far as the tubes, is completely invested.

As to the number and size of the tumours, rhinoscopy as a rule gives but uncertain information. On account of the great perspective shortening with which the parts under examination are represented, the lowest tumours only can be observed, while those lying above are either quite concealed or appear much smaller.

The pharyngeal tonsil, when hypertrophied, has as its basis a network of connective tissue, consisting sometimes of thick shining trabeculæ with narrow and roundish meshes, sometimes of thin fibres with wide angular meshes. Blood vessels, both arterial and venous, are very considerably developed. The main part of the adenoid vegetations consists of numerous round cells or lymph corpuscles. The epithelial covering is composed of ciliated cylindrical cells, sometimes of non-ciliated, less frequently of the pavement variety, the last being always found when a follicle approaches the surface.

Only a few years ago adenoid vegetations were considered as a comparatively rare affection, and were believed to occur only in certain countries, principally on the German coast, Denmark, Holland, and England. That this is not the case, but that it also is often observed in inland countries, is shown by the unanimous reports of many different observers; the reason being that, since the disease was made known, cases have been more diligently sought after.

Children appear to be specially predisposed to the disease. Many believe, and among them Bresgen, V. Lange, and Semon, that it is congenital, and that it is not observed till several years after birth, on account of its slow development. Without wishing to decide this question, I would like to call attention to the fact, that the predisposition of children to the affection does not seem so very remarkable, since other glandular organs, such as the tonsils and the lymphatic glands, are also in childhood very often affected with hyperplasia.

It has also been held that the presence of adenoid vegetations must be considered as an expression of a constitutional affection, mainly of scrofula. Although I cannot deny that among my patients scrofulous and anæmic children and adults were more numerously represented than the healthy and well nourished, still I must agree with those who look upon adenoid vegetations as an independent affection which often accompanies certain anomalies of the constitution. Chronic inflammation of the nose and pharynx, especially that accompanying measles, scarlet fever, and whooping cough, most frequently gives rise to this condition. According to Löwenberg, the condition also occurs in congenital cleft palate. All observers are agreed that the disease occurs most commonly between the ages of five and twenty; that is, it most frequently comes under treatment between these years. After twenty, the disease occurs but rarely. Along with it I have several times found nasal polypi.

One of the earliest and most frequent symptoms is alteration in speech. It loses its resonance, words sound muffled, and the nasal consonants cannot be pronounced. This way of speaking is designated "dead" and "toneless." In some cases it resembles the speech in acute anginæ, in others it is more like that in paralysis of the soft palate, which indeed has its function interfered with from the pressure of the vegetations and the hypertrophic condition with which it also is affected.

In cases which are pretty far advanced, breathing through the nose is interfered with, while in very severe ones it becomes altogether impossible. Parents often complain of the loud snoring of their children, by which those sleeping with them are awakened. The impossibility of breathing through the nose compels the patients to keep their mouths open, the under lip hangs down, and the countenance assumes an altered and stupid expression. Most children become pale and anæmic as the disease advances, their thorax remains flat, indrawn, and undeveloped, as we see in hypertrophy of the tonsils. That taste and smell also suffer, and that the removal of the secretion from the nose is impossible, is self-evident. Patients often complain of pain at the back of the head, or of a dull feeling of pressure in the head which they cannot clearly localise, and also of supra-orbital neuralgia, asthma, and other reflex neuroses, of which we shall come to speak more fully in discussing diseases of the nose.

Of very frequent occurrence are disturbances of hearing, from the slightest deafness and very indistinct tinnitus to well-marked difficulty in hearing and otitis media with perforation.

Increased secretion of mucus is one of the symptoms which is always present. The mucus, for the removal of which the most painful sneezing and hawking movements are made, is generally very thick and tough, viscid, and tinged with blood. The blood is either mixed with the secretion only in little streaks, or is brownish black. Sometimes it is fluid and comes away in large quantities, and sometimes the expectorations bear a striking resemblance to pulmonary sputa.

The course of chronic catarrh of the pharynx is very slow, especially when adenoid hypertrophy is also present. The vegetations may exist for many years without betraying their existence, until by the gradual narrowing of the naso-pharynx they excite the above-mentioned symptoms. Spontaneous recovery has never taken place as far as I know, but Lefferts mentions a case in which it did.

The prognosis of adenoid vegetations is favourable both as regards the removal of the affection and its complete cure without relapse, which up till now has never been observed. That restoration of hearing in cases of long standing, especially when perforation of the membrane has taken place, is not always possible, is very natural.

The diagnosis of chronic catarrh of the naso-pharynx and of adenoid vegetations must be guided on the one hand by the symptoms, and on the other by the condition as shown by rhinoscopy. As examination with the mirror is generally impracticable in children, the diagnosis must be confirmed by examining with the finger, by which method, too, one can become better acquainted with the number, size, and structure of the tumours. As an equivalent for the digital examination of children, which is always difficult and decidedly painful, Semon recommends that the permeability of the naso-pharynx be tried by injecting a small quantity of warm water by means of a ball syringe. If the water does not at once flow in a stream from the other nostril, but through the mouth, then it is certain that there is an obstruction in the naso-pharynx. This method is also recommended to prove whether or not operation has restored the passage. The suspicion that the case is one of hypertrophy of the pharyngeal tonsil will become a certainty, if children, in whom the nose and tonsils are unaltered, cannot breathe when their mouths are shut.

As the treatment of catarrh of the naso-pharynx, unaccompanied by hypertrophy of the adenoid layer of tissue, will be considered further on, let us now turn to the treatment of adenoid vegetations.

Treatment.—As a palliative to remove the extensive masses of mucus, the nasal douche with 1 per cent. solutions of chloride of sodium or bicarbonate of soda is best, while in complete obstruction of the passage to the pharynx some form of spray should be used. As nothing can be expected from internal remedies, surgical operations must be undertaken. The mildest of these is cauterising with solid nitrate of silver or solid chromic acid, which is to be melted on to a caustic holder, bent and twisted according to the situation of the vegetations. This method can only be used when the vegetations are soft and small, and must, in most cases, be continued at suitable intervals for several months.

Mechanical removal proves a much quicker and more radical cure. Although it is sometimes possible to make the growths disappear by crushing them off with the finger nail, still, as a rule, they require more energetic means.

The question at what age children should be operated on, must be answered by the condition of the child at the time. If disturbances of hearing, anæmia, and defective development of the thorax are present, an operation must be undertaken at once. But if the nose is still pervious, and there is neither deafness nor any disturbance of development, then one may wait till the child grows bigger, or till the symptoms increase; it is better, however, not to wait for the complete development of the disease, but to operate at once.

Another question is, whether or not it is necessary and possible to operate under control of the mirror.

The possibility of so doing must certainly be allowed; still I must affirm that, after many years' experience, it is exceptional for an operation to succeed when the mirror is used, and that it is not necessary to operate with it. If endo-laryngeal operations are brought forward as evidence in favour of the mirror, it shows that it has been altogether forgotten that the larynx is an organ which, both functionally and anatomically, cannot for a moment be compared with the pharynx. The only good, because the only practical point of view, and one which Michel, Schäffer, Lange, Semon, and others hold, is this, that operations can only be carried out without the mirror when, by previous inspection or palpation, the situation and extent of the vegetations have been carefully made out and observed. Whoever operates after taking such precautions, will never cut off the Eustachian prominence, or destroy a piece of the turbinated bones.

A still further question is, whether the operation should be done in one or in several sittings, and if anæsthetics should be employed.

The answer to these questions must be decided by the individual cases. In resolute adults, provided the vegetations are not too numerous, the operation may generally be completed at one sitting, and nearly always without an anæsthetic, even although the bleeding be somewhat considerable. If, however, the naso-pharynx is completely filled with the growths, then it is better to remove them at several sittings. With children the procedure is otherwise. Older and more sensible children, as a rule, submit to the first operation without much fuss. But if, as is often the case, the number of vegetations is very large, and a second application of the instrument, or a second seance, is necessary, then the resistance of the patient is in direct ratio to the persuasion and encouragement of the surgeon and the friends. Smaller and unreasonable children, who set themselves against every manipulation of the doctor, I only operate on when under the influence of an anæsthetic. The anæsthesia does not require to be very deep, only enough to interrupt excessive resistance, and last for a few minutes. On account of the danger of inspiring blood, mucus, or pieces of tissue, it is to be recommended here, as also when there are very many and large growths, to appoint a second sitting, or to operate with the patient's head hanging downwards.

As to the mode of operation, opinions differ very much. Some prefer to remove the vegetations through the nose, by means of ecraseurs, or the galvano-caustic snare. W. Mayer operates through the nose, by means of a ring knife, or an instrument like a lithotrite, guided by his fingers introduced through the mouth into the naso-pharynx. That this method through the nose is the shortest and most comfortable for adults, there is no doubt, and can only be objected to with right, on the ground that all the vegetations, especially those on the anterior superior part of the fornix, cannot be reached from the nose. Apart from the impossibility of passing suitable instruments through a narrow crooked nose, in which are also protuberances, it must not be forgotten that by far the greatest number to be operated on are children, in whom already, from anatomical reasons, the way of operating through the nose appears impracticable.

Therefore it is that many, and among them the author, prefer to operate through the mouth. Many recommend very warmly the wire snare, but as its introduction, on account of the bending of the wire, is very difficult, Hartmann has constructed a covered wire snare, which is preferable to the uncovered one. The galvano-caustic snare is hardly of any use, on account of the softness of the tumours, and to attempt to destroy numerous vegetations with the galvano-caustic cautery, is simply to prolong the operation *ad infinitum*.

We require, above all, a method which will put us in a position to remove as much as possible in the least possible time.

For this the sharp spoon principally employed by Justi, is suitable, and the cutting ring knife of V. Lange, or similarly constructed in-

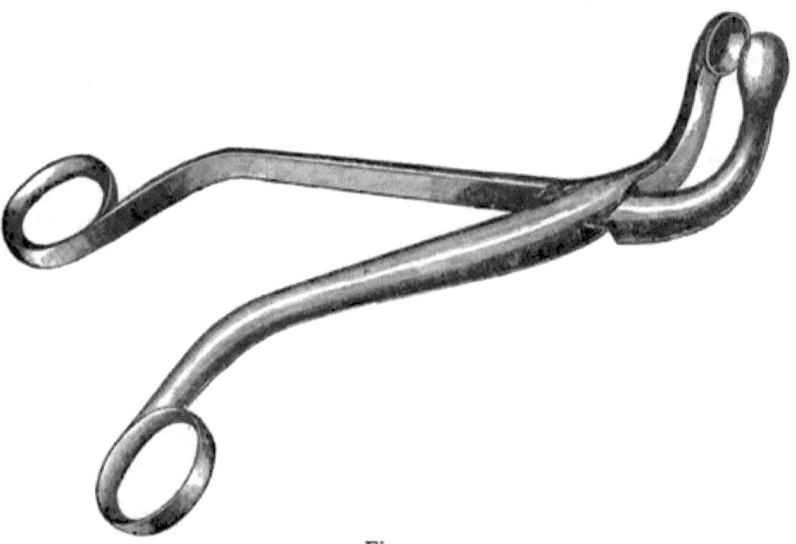

Fig. 3.
Schech's Post Nasal Forceps.

struments, curettes, &c. The author, who has used both instruments, both in flat and lamelliform vegetations, has every reason to be satisfied with them. When the vegetations are very numerous, large, and roundish, however, I make use of another method. Not from any desire to increase the number of instruments already in use, among which may be mentioned Delstanche's fixed adenotome, and

Bezold's lengthened metallic finger nail, but from real practical necessity, I constructed a post-nasal forceps, which in some degree resembles that of Stoerk, Löwenberg, and Solis Cohen. As represented in figure 3, it has a double curve, one at the handles, the other at the shoulder, concave anteriorly, so as to adapt itself to the naso-pharynx.

This forceps ends in two spoon-shaped blades, 14 millemetres long by 10 wide, and sharp all round, which come exactly into apposition when the handles are closed. By means of this instrument it is possible to reach every part of the naso-pharynx. If one wishes to remove vegetations from the posterior wall, then, the tongue being pressed down, the forceps can be directed to the desired spot by raising the handle; if necessary, it may be pushed as high as possible in a vertical direction, towards the fornix, and from there by slightly lowering the handle, on to the anterior surface toward the nasal fossæ, in order to remove the tumours situated there. A further advantage of this instrument lies in the possibility of crushing at one squeeze very large and tough pieces, in which the sharp spoon or the ring knife, as I have often experienced, would be caught, or on which they would make no impression. If the connective tissue base is very thick and hard, it may happen, as it does sometimes with ecraseurs, that the blades do not quite cut through it; as a rule, however, a few rotatory movements are sufficient to separate the tumour, and eventually sever the parts not quite divided. Let it also be observed that I generally always use the same instrument, whether operating on children or adults; and that it is only when the patients are very small, or the field of operations very narrow, that I use a thinner pair of forceps. The instrument may be obtained from Katsch, in Munich, the maker, at fourteen shillings per pair.

The bleeding following the use of this instrument, is less than that caused by the sharp spoon, and usually stops after a few minutes of itself, or when cold water is injected, containing, if necessary, some liquor ferri perchloridi. The pain is not very great, but patients often complain of shooting pains in the teeth and ears. After removal of the principal masses, the walls of the pharynx may be smoothed by means of the ring knife. Although it does not appear necessary to remove all the diseased tissue, after the experience we have had still it is better, as far as possible, to remove everything that can be reached, in case what is left behind should grow again.

The reaction following an operation completed at one sitting may be very severe; headache, fever, increased excretion of bloody purulent mucus, and general malaise, compel the patient to remain several days in bed. I have, however, never met with so violent a reaction since beginning to operate with the spoon forceps, and at several sittings; so that, in patients going about, or in merchants who travel a great deal, in weakly persons, and many children, the method is to be highly recommended. Even those children who are operated on under an anæsthetic, begin to play about in a few hours, or may attend school the next day. Injurious effects on the ear I have never seen. The reaction must be combated with applications of ice to the head, and sucking of ice. For after treatment, I use twice daily a 4 per cent. solution of boracic acid, for cleansing the naso-pharynx. Let one perform the operation with whatsoever instrument one will, and this is determined by practice, custom, or preference, the effect is always the same. When breathing through the nose is restored, the stupid expression disappears from the face, speech recovers its tone, and is distinct, the secretion of mucus stops, disturbances of hearing disappear, the pallid countenance becomes gradually healthy, and the thorax arches out. In some cases it is necessary, to hasten the restoration of hearing, to open the Eustachian tubes by forcing air through them.

Of all the divisions of the pharynx, the nasal part undoubtedly is most frequently diseased.

Chronic catarrh of this part, in its simplest form, manifests itself by uniform reddening and velvet-like swelling of the mucous membrane. The redness is sometimes so slight, that only an experienced eye is able to pronounce it to be a pathological condition, being in most cases only of a slight bluish red; in other cases the mucous membrane appears to a great extent mottled but smooth, or it may be slightly rough, owing to dilated tortuous blood vessels and swollen mucous glands, and covered by a gray greenish yellow secretion. The posterior wall of the pharynx sometimes appears divided into several flat or rounded longitudinal divisions, separated by shallow furrows; these are the cases in which already there is partial hypertrophy of the mucous membrane. Granular pharyngitis is the form which occurs most frequently. There are few affections which have been more discussed in late years than *granular pharyngitis*.

If one examines a considerable number of young and grown-up persons of both sexes, one may observe on the mucous membrane of the pharynx, but principally on that of the naso-pharynx and its immediate neighbourhood, extending longitudinally down the vertebral column, elongated or oval prominences of a deep or pale red or yellowish colour, as large as lentils, peas, or even beans. Sometimes there is but one, sometimes several scattered uniformly over the mucous membrane, sometimes they are in groups near each other, and often they form flat plaques of varying width. According to the examinations of Saalfeld and Roth, it has been conclusively proved that these granules represent a circumscribed proliferation of lymphatic tissue in the neighbourhood of the efferent ducts of the mucous glands. The efferent duct of the gland is dilated only in the region of the swollen tissue, the pavement epithelium of the mucous membrane extends over the granule, and appears either attenuated on its summit, or is completely absent. The mucous membrane between the granulations is sometimes normal, sometimes hyperæmic and intersected by dilated vessels.

Bresgen considers that the granulations are congenital. Roth contradicts him, and declares that he has never seen them in newly born children. In children under six they very seldom occur, but after six they gradually increase in frequency. Roth has repeatedly seen them develop first in adult life. My opinion is, that the more frequent occurrence of granulations along with increasing age, is explained by the intensity and frequency of the irritation which the mucous membrane of the pharynx meets with as age advances. While the development of the granulations is prevented in sucklings and young children by bland and unirritating food, and infrequent exposure to cold, they gradually develop more frequently however, as the children grow older, from their joining their parents at table, and also in consequence of unavoidable chills at school, as well as from the many intermediate children's diseases accompanied by acute angina. To these may be added, about the twelfth year or even much earlier, the irritation of cigars, and the taste for alcohol which increases towards adolescence, and still more towards manhood, and also the emergencies of vocation described in the introduction. What wonder is it that there is scarcely a single grown up male who is entirely free from granulations. Although females are not altogether spared, yet they do not suffer from them *en masse* as males do.

Combinations of the granular form, with simple chronic pharyngeal catarrh, come very frequently under observation. Here and there on the mucous membrane are seen prominences from the size of a millet to a lentil seed, having on their summit yellow spots, which are obstructed follicular glands filled with fatty material. They are also frequently met with on the soft palate and on the tonsils.

When the inflammatory process is localised to the lateral walls of the pharynx, it becomes known as *Pharyngitis lateralis hypertrophica*, or as inflammation of the lateral bands (Michel, M. Schmidt, Hering).

On inspecting the posterior wall, there is seen immediately behind the posterior pillar of the fauces a pale or intensely reddened elongated, round, or flat swelling, projecting toward the middle line, and extending upwards or downwards, or as is most often the case, in both directions. It is less clearly seen, or not seen at all, when the soft palate is slack, being best seen when the muscles of the pharynx are contracted. The size of this swelling varies from the thickness of a pencil to that of the little finger. As one may learn from inspection and palpation, it merges superiorly into the Eustachian prominence, representing therefore a true hypertrophy of the salpingo-pharyngeal fold. According to Zaufal, it may form an elongated fleshy swelling of the size of a hazel nut, becoming smaller inferiorly, and which may sometimes be compared to a conjunctival swelling with many granulations. In many cases the inflamed lateral band does not extend into the salpingo-pharyngeal fold, but remains entirely limited to the extreme lateral corner of the pharynx, and becomes lost below and external to the Eustachian prominence, or terminates in a slight curve anteriorly toward the hamular process. The Eustachian prominences and their folds, both at rest and in movement, are sharply differentiated from these bands, which, on speaking or swallowing, are subjected to a slight indrawing. The fold of mucous membrane in the neighbourhood of the posterior pillar of the fauces, when hypertrophied, very often becomes spread out into a flat and smooth or an uneven surface, consisting altogether of granulations, or beset here and there with single granules, which, by the movements which take place in choking or suffocation, is forced inwards. This thickening of the mucous membrane, which often resembles a string of beads, loses itself either in the seat of insertion of the posterior pillar of the fauces, or extends downwards, and gradually or suddenly changes back to normal mucous membrane at the level of the tip of the epiglottis.

Pharyngitis of the lateral walls is generally bilateral, though, as a rule, it is more strongly marked on one or other side. The mucous membrane lying between the hypertrophied lateral bands may be normal, beset with granulations, or changed in other ways. Hering differentiates pharyngitis lateralis granulosa from pharyngitis lateralis hypertrophica proper. By the former he understands a conglomeration of granulations, which are situated on the swelled or congested folds of the lateral wall round about the salpingo-pharyngeal fold, or underneath it; by the latter, the above-described hypertrophy of the salpingo-pharyngeal fold, and the parts of mucous membrane proceeding downwards from it. It is my opinion, however, that so strict a distinction cannot be justified.

Chronic inflammation of the soft palate generally occurs along with chronic catarrh of the posterior wall. In many cases it is the only change perceptible in the throat.

Its objective symptoms are reddening of the soft palate, which becomes roughened from dilated vessels and the swelling of numerous glands. It also looks thickened, the pillars of the fauces become thick red swellings, and their movement is more or less limited. The uvula, bluish red in colour, is rough and œdematous on its anterior surface from swelling of the follicular glands, and looks like a long thick and plump cone, which, with its point turned to one or other side, hangs down on to the dorsum of the tongue, or sticks to the posterior wall. The mucous membrane of the uvula, principally at its most dependent part, hangs down like an empty sack. Sometimes the epithelium looks dim and milky.

Hypertrophy of the tonsils must also be considered as a result of chronic inflammation.

Acute phlegmonous and diphtheritic anginæ taken etiologically must first come under consideration. Sometimes an intra-tonsillar abscess does not burst, but runs a chronic course. Children show a peculiar predisposition to the affection—especially the anæmic and scrofulous; in some cases the predisposition seems to be hereditary. Gerhardt believes that it often appears as the last evidence of syphilis in the parents. Morell Mackenzie considers that its occurrence in the first months of life is due to eczema and impetigo of the face and head, or to blennorrhœa neonatorum and blennorrhœic rhinitis, while its appearance at puberty is traced back by many to a sympathetic connection with the genital organs.

Hypertrophy attacks all the tissues of the tonsils. The follicles are enlarged and increased, the interstitial connective tissue thickened, and also the mucous membrane covering them. The consistence of the enlarged tonsils varies; in scrofulous children they are soft and elastic; in the cases which so often occur after acute anginæ they are generally very hard, and in many cases chalky deposits are present.

On inspection one is struck with the increase in size of one or both tonsils. The enlargement takes place in all directions, causing the organs to reach far into the isthmus of the fauces. In very marked cases they are in close apposition, or leave only a small space between them at their upper part, in which the uvula, which is often hypertrophied also, is seen, while the posterior wall can hardly be distinguished, if at all. The tonsils reach upwards and backwards into the naso-pharynx, downwards to the base of the tongue, and force the soft palate forwards into the cavity of the mouth. Their mucous membrane is either reddened or pale, sometimes of a yellowish colour, its surface is generally rough and covered with retiform scars and shallow depressions, the parenchyma, cleft, ragged, and in some places atrophic or wanting. Their hypertrophy is very often combined with that of the pharyngeal tonsil, with granular pharyngitis or with hypertrophy of the lateral bands. As hypertrophy of the tonsils commonly arises from inflammation, there is generally also in consequence indolent swelling of the lymphatic glands of the neck.

The pharyngo-laryngeal cavity may be affected by itself, or may participate in affections of the naso-pharynx.

The objective condition here is quite simple: the posterior wall of the pharynx, as seen with the laryngeal mirror, is more or less reddened, covered with mucus, rough and slightly nodular. This roughness is almost entirely caused by the swelling of the mucous glands, seldom by granulations, which are distinctly differentiated from the former by their shape and larger circumference. In diseases of the pharynx the larynx always shares. The aryepiglottic ligaments, the interarytenoid commissure, sometimes the whole posterior wall, appear more or less reddened and swollen, the mucous membrane of the interarytenoid region arched forwards, and covered with thick mucus. Less frequently the redness extends to the vocal cords, and, as a rule, the closure of the glottis is incomplete in consequence of the swelling of the mucous membrane, and from paresis of the muscles.

Having now seen that chronic pharyngitis is most frequently accompanied by hypertrophy of the mucous membrane, we must also consider what takes place in the event of atrophy.

This form is called *pharyngitis chronica atrophicans sive sicca*, or "rarifying" catarrh of the pharynx. Atrophy attacks the mucous membrane in varying extent, and in different degrees. It must be carefully remembered that there are very often hypertrophied patches near those already atrophied. Although it cannot be definitely proved that atrophy always develops from hypertrophy, because one is seldom in a position to observe chronic catarrh of the pharynx from first to last, nevertheless this appears probable by analogy. The question whether atrophy can develop primarily, *i.e.*, without previous hypertrophy, must be looked on as an open one. The pharyngo-nasal and pharyngo-oral cavities are undoubtedly the parts of the pharynx most often affected by atrophy, which very often is sharply defined at the level of the uvula, or at the seat of its insertion on the soft palate.

The mucous membrane of the posterior wall appears thin, delicate, and glistening, and on the contraction of the muscles shows fine distinct light reflecting folds. The constrictor of the pharynx, which can be seen under the mucous membrane, gives it the appearance of raw smoked flesh, of a purple red or pale colour. The cavity of the pharynx looks wider, when atrophy has affected large portions, but especially so when the adenoid tissue of the naso-pharynx has disappeared. The Eustachian prominences and the posterior parts of the turbinated bodies, are seen very distinctly, partly in consequence of atrophy of their mucous coverings, and partly from that of the tissues in their immediate neighbourhood. In this atrophy the uvula often participates, when it hangs down relaxed, as a thin needle-like process. In consequence of the extinction of many mucous glands, secretion is more or less diminished and changed. The secretion covers the mucous membrane like a carpet, and imparts to it a varnish-like gloss; it always adheres very firmly to the surface beneath it, on account of its being so devoid of water. So much so that one has to use a pair of forceps to remove it. The assertion, that dry catarrh of the pharynx only arises in consequence of fetid rhinitis, and that the fetid secretion, which often dries up and forms hard scabs, always comes from the nose, I must deny in the most decided manner. Although I have convinced myself that fetid rhinitis is invariably accompanied by pharyngitis sicca, yet in many cases of pharyngitis sicca, the nose is not affected at all.

Dry pharyngeal catarrh occurs in children as well as in adults; it cannot be contradicted, however, that it occurs oftener in more advanced age, or in otherwise badly nourished persons, *e.g.*, the scrofulous, anæmic, and tubercular. It seems to stand in a certain not yet more closely discovered causal relation to diabetes mellitus, and Bright's disease. My attention has been repeatedly called to the presence of these affections in atrophic pharyngeal catarrh, and I recommend, therefore, that in every case the urine be examined.

It only remains now to consider that form of chronic pharyngeal catarrh, which Hoppe has called "fetid stinking pharyngitis." According to the latest observations, the process appears to be identical with fetid rhinitis. Roth made certain, by plugging the nasal fossæ, that the fetid secretion does not originate from the nose, but is produced from the mucous membrane of the pharynx itself. The affection occurs sometimes in combination with ozæna, sometimes by itself. It appears to be very nearly related to Stoerk's blennorrhœa, of which, however, it is characteristic, that the process always begins in the nose first, and spreads as it runs its course, to the pharynx, larynx, and trachea.

Let us now turn to the symptomatology of chronic pharyngeal catarrh.

Among the earliest symptoms are perverted sensations in the pharynx.

Dryness of the throat is a symptom very often complained of, although phlegm also is not less frequently a cause of complaint. One would be very much deceived were one to believe, that in the former case the mucous membrane was without exception dry, and covered with but scanty secretion, while in the latter case, with a copious thin secretion. It is only in pharyngitis sicca that dryness is always complained of. The quantity of secretion varies just as much as its quality. The secretion is, as a rule, grey and viscid; sometimes it is more like mucus, or is muco-purulent and streaked with blood, but very seldom is it quite bloody. The secretion, when yellow, purulent, and fluid, or dried up into scabs, is generally a sign of ulceration or ozæna. It sticks to the mucous membrane, either on isolated spots in the form of small grey lumps, or, as in atrophic catarrh, covers the whole of it uniformly.

The natural consequences of these anomalies of secretion, is the necessity of removing the secretion, real or imaginary, by hawking. The removal is accomplished either by deep and sounding expiration, the so-called "hem hem," or by hastily drawing air into the naso-pharynx, with the mouth shut, or by simultaneous violent quiverings of the soft palate. Another cause of frequent hawking is from the patient feeling as if he had a foreign body, a hair, a bone, a needle, a grain of sand, or some such thing in his throat. Through fruitless and repeated attempts to remove this foreign body, the reddening and sensitiveness of the mucous membrane become greater, and hence vomiting and choking, especially in the morning, take place, and increased necessity for swallowing arises. This last, according to Rühle, often leads to abnormal inflation of the stomach, furring of the tongue, loss of appetite, and indigestion. There is seldom any real pain in swallowing, but darting pains may be occasioned by dried up masses of secretion, especially in the atrophic form.

Unpleasant burning, sometimes cold, sensations, itching and oppressive feelings, occur spontaneously, or on taking food, also on smoking or partaking of hot highly seasoned dishes, and particularly alcohol. Many patients experience abnormal sensations, even on taking pure fresh water; in the atrophic form the symptoms are very often mitigated after taking irritating food, apparently in consequence of the secretion being increased for the time by the augmented blood supply; sometimes for several hours they altogether disappear. The fact that many persons affected with numerous granulations and hypertrophy of the lateral bands, or even with pharyngitis sicca, experience no bad effects, while others display the most alarming symptoms on the appearance of the slightest change, depends partly on the varying excitability of the nervous system, partly on the accompanying inflammatory symptoms. M. Schmidt explains the absence of symptoms in spite of the lateral bands being inflamed, by the circumstance that the latter do not reach as far as the superior constrictor of the pharynx, and are not disturbed by the contraction of this muscle.

In highly susceptible, but also in robust persons, neuralgic symptoms very often appear in distant organs. In women, reflex spasms, according to my own and Sommerbrodt's experience, occur pretty frequently at the entrance of the œsophagus. Asthma, megrim, and other symptoms, which will be discussed more fully under affections of the nose, may occur as consequences of ordinary pharyngeal catarrh. Thanks are due to Hack for having called attention to the significance of these symptoms, and to the spontaneous occurrence of sudden darting pains between the shoulder blades, and rheumatic pains over the clavicles, usually occurring only on muscular contraction.

Another symptom to be mentioned here is cough. Although in far the greatest number of cases no cough is observed in chronic pharyngitis, and when present is confounded with hawking, yet now, by the careful observations of Sommerbrodt, Spamer, and others, it has been undeniably proved, that cough may really be excited by the condition of the mucous membrane of the pharynx. I myself have seen cases of individuals in whom cough has arisen spontaneously, and also cases in which it has been excited with remarkable promptness by touching the granulations and lateral bands. The term "pharyngeal cough" has been decidedly justified, apart from the positive experiments of Kohts. Stoerk explains it by irritation of the larynx from flowing down of the secretion, but in this he is only partly right, for when the larynx is also affected, however slightly, the occurrence of cough requires no further explanation.

Chronic catarrh of the pharynx has a most injurious effect on the voice.

Be the patient teacher or officer, singer or actor, all complain of the voice soon becoming tired, of the loss of metallic ring, and of the difficulty of producing high notes. These symptoms are explained partly by the friction and irritation of the larynx in consequence of constant hawking, partly by anatomical changes which the mucous membrane of the pharynx has undergone. Michel, with whom I entirely agree, explains them in the following manner :—

The pharyngeal wall forms the principal reflector for the waves of sound streaming out of the glottis; here and against the soft palate they strike first. If this wall be not smooth, but is rough and uneven from granulations and hypertrophies, dull sounds must be formed, just as an uneven mirror can only give a distorted image of an object before it. By being badly reflected, therefore, the sound suffers damage, it becomes weakened, it does not carry far, or what is the same thing, it loses its metallic ring. Granulations and superficial hypertrophies, situated on the posterior pillars of the fauces, load them, make them stiff, and more or less rigid, whereby their mobility is interfered with; thus they become incapable of vibration, do not transmit the waves of sound which strike them, and hence a part of the sound is again lost. Pharyngitis lateralis hypertrophica hinders the movement of the soft palate, and prevents its rising up, so that it acts as a mass of flesh, and further damages the resonance. Another hindrance to the development of the voice, lies in the swelling, heat and dryness of the mucous membrane, occasioned by continued loud speaking and singing. Through the attempt to overcome these various hindrances, and to supply the deficiency of tone, the voice is often unduly overstrained, so that there follows in consequence morbid tiring of the laryngeal muscles, weakness of the voice and loss of metallic ring. Finally, one must not forget that according to Gerhardt's observations, paralysis of the vocal cords may arise reflexly after morbid processes in the pharynx.

Looking at the sum of these different disturbances, it is not to be wondered at, that whenever chronic catarrh of the pharynx influences the general health very specially, it also has a very injurious effect upon the patient's mental condition.

Depressed disposition of mind, thoughts of dying, the idea that incurable consumption or cancer of the throat has developed, lead through defective nourishment to restless insufficient sleep, lassitude, anæmia, weariness of life, with thoughts of suicide, and are signs of real throat hypochondriasis. Anxiety about his life causes the patient to choose most carefully his food and liquors, to avoid all, even the weakest alcoholic drinks, and even all solid nourishment, to use the most nauseous medicines, to wrap the thickest scarfs round his throat, and to shun even to fanaticism fresh air and contact with cold water. How far this psychical aberration may go, may be illustrated by the case of a lady, one of my patients, who for months would eat nothing but new laid eggs, which had to be most carefully examined, sometimes even with the microscope, because she had the fixed idea, that the itching sensations which tormented her, were caused by the mixing of sand and stones with her food.

In conclusion it remains now to discuss the symptoms which are occasioned by hypertrophy of the uvula and tonsils.

As the most terrible consequences have been attributed to hypertrophy of the uvula, it does not appear unnecessary to separate the real from imaginary disorders.

That enlargement of the uvula may be the cause of pulmonary consumption is now-a-days scarcely believed by any old woman. That it may cause symptoms of suffocation by hanging down on to the vocal cords, also belongs to the domain of fable. It does happen, though very rarely, that the uvula reaches down as far as the epiglottis, and gives rise to an irritating cough, and under certain circumstances may cause spasm of the glottis. Generally, however, the symptoms are confined to the feeling as of foreign bodies, to tickling, and phlegm—a suffocative sensation and inclination to vomit. The voice is often thick on account of paresis of the azygos uvulæ and nasal from overburdening of the soft palate.

The symptoms of hypertrophy of the tonsils are almost identical with those of hypertrophy of the pharyngeal tonsil.

Respiration is in this case loudly audible, both when the patient is moving about or when resting. Sleep is restless and interrupted by sudden attacks of choking, and very disturbing on account of snoring. The mouth being open, the face here assumes the well known dull expression, and sometimes there are disturbances of hearing. Speech, through narrowing of the conducting tube, sounds weak, muffled, and toneless, often with nasal timbre; the compass of the voice in singing is limited, and a raising of the pitch of the voice is said to have been often observed. There is no pain on swallowing, but deglutition is rendered difficult and slow for solid, hard substances and for large pieces. Complaints of feeling foreign bodies, increased secretion of mucus, increased hawking, and increased inclination to swallow, are the same as in the other forms of pharyngeal catarrh. Hypertrophy of the tonsils has an especially injurious effect on the development of children, which is generally far more pronounced than that of adenoid vegetations, which has been already mentioned.

The face and nose do not grow in proportion to the rest of the body, the nostrils remain small, the upper lip becomes larger, the nasolabial folds are flattened, the roof of the mouth is strongly arched and pointed, the alveolar processes become small, and the incisor teeth overlap one another. In consequence of a deficient supply of oxygen, the children become anæmic, the expression dull, the face puffy and cyanotic, and the muscles badly developed. When the disease is of long duration, the thorax becomes affected. It is flat, long and small, in a word, paralytic; often assumes the rachitic character of pigeon breast, or shews on the lower part of the sides the impression of the arms.

The diagnosis of chronic catarrh of the pharynx depends on the subjective and objective symptoms. Let the beginner always bear in mind, that in very many cases, the objective condition bears no ratio whatever to the subjective symptoms, so that he may not make a mistake in his diagnosis. Only when objective symptoms are altogether awanting, or are very insignificant, can the diagnosis "nervous affection of the throat" be made. Since ulcers never appear in idiopathic chronic catarrh of the naso-pharynx, when they are seen, one immediately thinks of syphilis, tuberculosis, lupus, herpes, &c. The differential diagnosis between catarrh of the larynx and bronchi and pharyngitis is as a rule quite easy, but sometimes requires prolonged observation to make sure. On the assertions of patients that pain or unpleasant sensations are only on one side, or that they are situated high up, in the middle, or deep down, much reliance cannot be placed, on account of the small sense of localization in the throat. Yet, in some cases, I have found touching diseased or suspected parts of the mucous membrane, especially granulations and the lateral bands, with a sound, to be in so far a diagnostic expedient as that it makes the patients, who are altogether unprepared for it, almost invariably complain of sudden pain, and point out with certainty, as the situation of their troubles, the part of the mucous membrane that has been touched.

The disease may run a course of many years, or may even last the whole of one's life. The symptoms may be at one time less, at another more pronounced. The condition of the mucous membrane may for a long time remain entirely unchanged, but generally hypertrophy steadily increases, or atrophy sets in.

The prognosis is favourable as regards life. Although chronic catarrh never causes death directly, still it may thoroughly embitter the lives of those affected by it, as may be concluded from the described complaints. While ten years ago chronic pharyngeal catarrh was thought to be incurable, now, thanks to the late advances of therapeutics, it has become otherwise. The hypertrophic form may be completely and permanently cured; the atrophic form alone has a bad prognosis, since as yet we have no remedy which can restore an atrophied mucous membrane to the normal condition.

Treatment.—There are few affections about which so many undecided opinions as to treatment exist as chronic pharyngeal catarrh. If I therefore describe its treatment in detail, I am induced to do so by the conviction that such a description is very much needed; but the patients who are really anxious to get rid of their troubles, must never forget that cure is only possible with the greatest care and perseverance, and that an affection which has existed for years cannot be removed in a few weeks.

As one can hardly ever satisfy the *indicatio causalis*, one must in most cases be satisfied with a prophylactic and hygienic-dietetic method. It is quite evident that anæmia, scrofula, and hæmorrhoidal conditions must be removed, the partaking of strong alcohol, highly seasoned or too hot foods, smoking and snuffing forbidden, or else limited as much as possible. Workers in dusty atmospheres must protect themselves by wearing respirators, and while under treatment avoid their usual places of business; professional speakers must take rest and use their voices sparingly.

Since chronic pharyngeal catarrh represents a local malady, its treatment must also be principally local. It is therefore evident that baths, mineral waters, or hydropathic cures, except in cases due to plethora, too high living, and hæmorrhoids, are not the proper remedies. Nevertheless experience teaches, that the strong belief which physicians and the public still have in the healing power of certain springs, promote yearly a pecuniary sacrifice which bears no proportion to the results obtained. From the unusual prevalence of chronic pharyngeal catarrh, it is not to be wondered at, that chronic pharyngitis has found a place among the indications of most health resorts, and hence it is that both alkaline and saline springs, and among the latter especially those containing iodine, and also sulphur springs and earthy mineral waters, are recommended for their great curative power. The use of all these waters consists solely and only in this, that the patients, during the cure, trace to them an alleviation and seeming improvement of their subjective complaints, which, however, after their return home reappear with all their former activity. Both Bresgen and Lichtenstern have repeatedly but unsuccessfully urged the uselessness of these water cures; the baths are too convenient as places to which everlastingly querulous patients may from time to time remove, for one to wish to renounce them.

Of the same value as baths are gargles, inhalations, and applications by sprays. The former, as previously mentioned, are only useful in diseases of the isthmus faucium; the latter are of value for cleansing and moistening the mucous membrane, but have no healing power. I prefer the cold spray, partly because it more quickly blunts the excessive irritability, and partly because it renders the mucous membrane, already all too much accustomed to heat, gradually more capable of resistance. I only use solvents, and principally 1—2 per cent. solutions of bicarbonate of soda, carbonate of potash, and chlorate of soda; in some cases a 4 per cent. solution of bromide of potash for soothing. For cleansing the naso-pharynx I use either the nasal douche or the above-named solutions, simply snuffed up through the nose, or, in the case of children, introduced with a spoon.

In painting and insufflation, we possess, on the other hand, rational means of applying remedies to the pharynx.

If treatment is to be applied for uniform reddening, with velvet-like swelling of the mucous membrane, or for recent cases of hypertrophy of the uvula, one should begin with daily paintings with a strong solution of tannin (5 to 20), which later on may be replaced by solutions of nitrate of silver, gradually increasing in strength (grs. 25–50 ad ʒi).

When there is considerable swelling, no remedy is so good as Mandl's solution of iodine. It is a good thing to have several solutions of varying strength, to be used according to the severity of the case. Mandl added carbolic acid to his solution for producing anæsthesia of the mucous membranes. I prefer oleum menthæ piperitæ, on account of its better taste. The three different strengths of the solution are as follows:—

	1.	2.	3.
℞ Potass. Iodid.	grs. 25	grs. 50	grs. 75
Iodi puri	,, 6¼	,, 12½	,, 20
Glycerini	ʒi	ʒi	ʒi
Ol. Menth. Pip. guttas tres.			

Bresgen uses still stronger solutions, although the above are sufficient for general use; the weakest of them at first causes violent burning, which, however, generally stops in a few minutes. The painting, which must reach every diseased part, must be done with a bent brush when the affection is in the nasal and laryngeal parts of the pharynx, at first daily, but afterwards, when the stronger solutions are used, every third day.

Treatment must be continued according to the degree of swelling, and extends generally over from four to six weeks. For insufflation, a mode of treatment which is, in my opinion, much less effective, and which an intelligent patient can apply himself, the best remedy is nitrate of silver (1 part to 20 of starch). Morell Mackenzie recommends catechu in powder (3 grains in each dose), also persulphate of iron (1 part to 3 of starch), or resin of eucalyptus (1 part to 3 of starch).

Although one may succeed in removing smaller granulations by applying a solution of iodine, larger ones are not at all or only slightly influenced by it. Single large granulations may for a time be got rid of by crossed scarification, followed by painting, but for their radical cure, as well as that of hypertrophied lateral bands, there is only one remedy, and that is the galvano-caustic point, which has been introduced into practice by Michel.

Nothing could be said against the use of Paquelin's thermo-cautery, were it not for the fear patients have at a glowing hot instrument being introduced into their mouth. The mechanical removal by means of sharp spoons or curettes, and the chemical destruction by means of solid nitrate of silver, chromic acid, or London paste, have no advantages, but on the other hand many disadvantages.

For cauterising granulations one uses either a galvanic cautery ending in a small round plate, or when the granulations are larger and flat, as well as for destroying the lateral bands, a galvano-caustic knife, bent transversely at a right angle, according to the size of the surface to be cauterised, which, the tongue being pressed down, is to be placed cold on the spot to be touched, and then brought to a glowing heat. It is best not to allow the instrument to come to a white heat, which is prevented most easily by interrupting the current, and it is also recommended that the firmly adherent cautery be removed while still hot. When there are only small granulations present, it is best to destroy them all at once, but when the granulations are numerous, it is necessary, on account of the reaction, to perform a second or even a third operation.

Very great caution and practice are required for cauterising the lateral bands. When at rest they are often so completely concealed behind the posterior pillars of the fauces, that it seems advantageous to cauterise them during phonation or retching, when they approach the middle line. The greatest care must be taken to avoid touching the posterior pillar of the fauces or other healthy parts. In hypertrophy of the salpingo-pharyngeal folds, I allow the patient to hold down the tongue himself with the bent spatula, while I, in order to reach as high up as possible, and to avoid injuries to the soft palate, push up the latter with some suitable instrument, and cauterise the part from above downwards in long streaks. When the lateral bands are very hard, as they often are, and when the granulations are situated laterally, complete destruction of all diseased parts can only be accomplished by very thorough cauterisation.

While cauterisation of the granules, when situated in the middle, generally gives rise to but little pain and slight reaction, yet when they are situated laterally, and especially when the lateral bands are to be cauterised, the operation is unusually painful. When, therefore, the disease is bilateral, it is scarcely possible to complete the operation at one sitting. The reaction is almost always very great; the difficulty in swallowing next day, or even after a few hours, reaches such a height that the patients can only take liquid nourishment, and even refuse that, so that they may not have to swallow. The violent pain is caused by, not only the extraordinary abundance of nerves in the lateral walls, but also the disturbance of the inflamed parts by the contraction of the constrictor of the pharynx. The reaction is seldom so violent as to compel patients to remain in bed for some days from headache, fever, or general discomfort.

If, however, such be the case, ice must be applied to the head, and also sucked, and gargles be used, as cold as possible. When the symptoms are not so severe, the pain gradually disappears after from three to five days, and only slight uneasiness or flying pains remain during swallowing. The eschar, which forms after cauterising, and resembles very much the exudation of croup, is shed generally in from eight to fourteen days, sometimes longer. Whether a second sitting is necessary can only be decided when the resulting imflammation and swelling have completely gone down, when the destruction of the remaining diseased parts, or of the other lateral band, may be proceeded with. When reaction is slight, destruction of the granulations may be proceeded with earlier.

However easy and convenient the galvano-caustic method may be, still, besides being very painful, it has another disadvantage, which is, that it takes up a great deal of time, and this is a point which, with patients from a distance, and who have neither time nor desire to await reaction or make repeated journeys, must be considered. I have, therefore, in cases where the lateral bands were very hard and thick, often excised them with a knife.

This operation succeeds, however, only in patients who have been anæsthetised, or otherwise made insensible. The pain during the operation is much less than afterwards, and the bleeding scarcely worth mentioning. An injury to the carotid is possible when the cautery is at a white heat, but only through carelessness or tota ignorance on the part of the operator. The cicatrix is in both cases very slight, scarcely visible, and only causes discomfort when the posterior pillar of the fauces has also been cauterised, causing it to adhere to the pharyngeal wall.

Should hypertrophy of the uvula not be checked by the above mentioned modes of cauterising, then its amputation comes to be considered. It seems necessary to lay down very minutely the indication for cutting off the uvula, on account of its being practised now-a-days to cure all possible diseases. It is, as Semon also points out, only indicated when the uvula really hangs down to the larynx, thereby giving rise to cough and spasmodic respiration: when other symptoms are caused in the pharynx, entirely due to its enlargement; when it hinders operation on endo-laryngeal tumours, the opening of abscesses or local treatment; and lastly, when it is the seat of malignant neoplasms.

I perform the operation with Stoerk's guillotine; the simplest way, however, is to use a knife or scissors. The whole of the uvula ought not to be removed, a piece, at least one centimetre long, should be left. The wound heals in a few days without much pain, still it is necessary to pay some attention to the choice and temperature of the diet, and it is also advisable to order antiseptic gargles.

For removing hypertrophy of the tonsils, the most diverse remedies have been proposed. Apart from the above mentioned pigments, which are expected to do good only in recent cases, some have hoped to produce diminution, by rubbing with crystals of alum with or without massage, by inserting zinc, copper, and nitrate of silver darts, or iodide of potash crystals, or by parenchymatous injection of iodine, acetic or chromic acid, and similar materials. Apart from the trouble and pain they give, these methods are entirely useless.

Diminution by means of the galvanic cautery appears more rational, only the slowness of the method and the necessity of repeated applications, especially in children, makes it at the same time appear doubtful.

The only really rational remedy is amputation, by means of cutting instruments. That the tonsillotome is preferable to the knife, cannot, in my opinion, be doubted for a moment, but it is hard to say which tonsillotome is best, since custom and practice carry so much weight.

I use a tonsillotome of an English pattern, having two lateral hooks, which, when the knife is drawn back, strike inwards and seize the amputated piece; the instrument is pushed over the tonsil just as far as it is wished to remove, hence it is a good plan to have several instruments of different sizes.

To remove the whole tonsil is not only unnecessary, but also absolutely objectionable, on account of the danger of fatal hæmorrhage from the carotid; only the part projecting beyond the pillars of the fauces should be amputated. That by removal of the tonsils the sexual function is interfered with is a fable, as is also the theory held by some, that the voices of actors, singers, and speakers, undergo a disadvantageous change. Many object to tonsillotomy on account of the bleeding. Fatal hæmorrhage through injuring the carotid is impossible if the tonsillotome be used; with the knife, however, it may occur. If the carotid is once opened into, ligature is the only remedy. Smart bleeding, according to Lefferts, is best controlled by torsion of the vessels or by digital compression; it arises generally from dilated veins or the artery in the anterior pillar of the fauces. Moderate bleeding, which is most common, stops in a few minutes of its own accord, or after gargling with iced water; it may also be stopped by applying some chloride of iron solution.

Another danger, which deters many from tonsillotomy, is diphtheria. That it has had its victims is indeed true, but such occurrences are very rare. Still tonsillotomy, when it is possible, should be done in a healthy private dwelling-house and not in a hospital, and should never be performed during an epidemic of diphtheria.

Lastly, the proposal has been made to amputate the tonsils with the galvano-caustic snare. From my own experience, I must speak against this method of operation, on account of the trouble the proceeding involves, and the difficulty of obtaining an even cut surface, and further, because the danger of hæmorrhage, which it is intended to prevent, is not at all excluded.

As regards the treatment of pharyngitis sicca, it must be limited to careful cleansing and moistening of the mucous membrane several times a day with simple remedies, *e.g.*, warm milk, mucilaginous decoctions, weak glycerine solutions, one per cent. soda or common salt solutions. Astringents only do harm; but on the other hand, the malady may be temporarily diminished, or even cured, by painting with strong solutions of iodine in glycerine.

The treatment of pharyngitis fœtida and of Stoerk's blennorrhœa must be conducted according to the principles laid down in diseases of the nose.

PHLEGMONOUS PHARYNGITIS.

Synonym.—Pharyngitis phlegmonosa, &c.

When inflammation extends from the mucous membrane into the submucous tissue, or the parenchyma of the organ, then phlegmonous or parenchymatous pharyngitis is spoken of; or again, because the phlegmon is accompanied by œdema, and often ends in the formation of pus, the lesion is named œdematous, or suppurating, pharyngitis

Phlegmonous pharyngitis may, just as in the case of catarrhal pharyngitis, be caused primarily by chills and other mechanical or thermal influences, by operative procedures, and by cauterisation and galvano-caustic manipulations. There are many persons who are predisposed in a remarkable manner to pharyngeal phlegmon. Sometimes it occurs at the conclusion of erysipelas of the face, and passes away with vesicular formation on the cutis. As a secondary affection it most usually accompanies acute infectious diseases, and especially scarlet fever. Not uncommonly it is of toxic origin, or a result of swallowing mineral acids, caustic alkalies, broken off pieces of lunar caustic, &c., and is combined with sloughing of the mucous membrane. That the predisposition to phlegmonous inflammation is increased by severe general disorders which interfere with nutrition, is not improbable, at least, the case of a patient of Eberth's, affected with bulbar paralysis, in some degree supports this idea.

Phlegmon may attack any part of the pharynx, preferably, however, the soft palate, the tonsils, and the peritonsillar connective tissue.

The disease is developed either from gradual exacerbation of acute catarrh, or suddenly with the most severe general symptoms, such as chill, fever (as high as 104° F.), quickened pulse, headache, exhaustion, and great pain on swallowing. The sensation of dryness experienced at first, is soon replaced by the secretion of thick tough mucus, and later, by copious salivation; the mouth also participates in the inflammation, the tongue becomes thickly furred, and there is inclination to vomit, loss of appetite, and fœtor of the breath. As the disease advances, the pain on swallowing increases to an unbearable degree, so much so that it is declared that, next to tuberculosis, phlegmon is the most painful of all pharyngeal affections. On account of fever, pain, and sleeplessness, and the nourishment that can be taken being reduced to a minimum, even robust patients come to be very much reduced. Food is either forced down with great difficulty, distortion of the countenance, and expressions of the most violent pain, or, when there is paralysis of the soft palate, is at once thrown up again through the nose and mouth. Speech is very much impeded, deficient in timbre, and nasal; the mouth, in consequence of infiltration of the bucco-pharyngeal fascia, can hardly be opened, and hence inspection and medication are rendered most difficult. Hoarseness only occurs when the larynx is also affected, and difficulty of breathing arises, either from narrowing of the isthmus of the fauces, or from œdematous laryngitis.

At a still later stage, the exudation from the submucous tissue is either reabsorbed, or changes into pus as the disorder increases. If the pus is not removed artificially, it may become evacuated during sleep, or when hawking or gargling. The quantity is often very considerable; it is thick, mixed with blood and shreds of connective tissue, and very fetid. After the pus is evacuated, the condition of the disease is suddenly changed; the violent pain is almost gone, or becomes bearable, the breathing becomes freer, swallowing is again possible, and refreshing sleep is procured.

The objective symptoms of phlegmonous inflammation of the naso-pharynx, consist in considerable redness and swelling, especially of the adenoid tissue and the Eustachian prominences. It is generally so great as to completely close the lumen of the naso-pharynx, and to make respiration through the nose impossible. The secretion is purulent, sometimes bloody, hæmorrhages into the mucous membrane often appear, and the middle ear very frequently participates in the affection.

On the soft palate phlegmon is usually unilateral. On the affected side it appears of a deep red colour, is somewhat thickened, and projects into the cavity of the mouth, most prominently at the edge of the hard palate. Its contractility is at the same time interfered with, or it may become quite immoveable. The uvula appears as an œdematous bluish red sack; the pillars of the fauces are thickened, sometimes more especially the anterior, so that with the adjacent parts it projects into the mouth, and at others the posterior, which then lies on the posterior wall of the pharynx—an œdematous swelling of the thickness of a finger. The phlegmon may stop here, or it may extend along the posterior wall, and alter it into a swollen surface, narrowing the lumen of the throat. Thence it may spread to the larynx, generally on the side corresponding to the affected half of the palate, more seldom on both. The aryepiglottic ligaments, or the epiglottis, including the ventricular bands, are reddened and œdematous, the lumen of the larynx is more or less narrowed, and when the phlegmon is bilateral, is almost entirely obstructed.

If the inflammation goes on to suppuration, there is formed at one spot, as the swelling and redness increase, a prominence which points out the seat of the abscess. In diffuse swelling there is generally no fixed point where pus is to be found.

In phlegmon of the soft palate the tonsils may remain completely intact, may become catarrhal, or may be involved in the phlegmonous process. Very often they are attacked primarily, either one, or both, one after the other. The affected tonsil becomes increased in size, and pushes the anterior pillar of the fauces before it, at the same time forcing the posterior against the posterior wall of the pharynx; it also pushes the uvula to the opposite side, and narrows the isthmus of the fauces. When both tonsils are affected, the uvula gets fixed between them, or is forced forwards and upwards; the isthmus is narrowed to a small slit, or completely closed; at the same time there is often suppuration of the follicles. Symptoms, both subjective and objective, are most intense, when both the tonsils and the soft palate are attacked together. If the affection progresses to a cure, the tonsil becomes gradually less; but if an abscess form the swelling generally increases to some extent, without, however, it being possible to recognise the seat of the pus with the eye.

If the connective tissue between the tonsils and the soft palate is affected, the disease is known as *peritonsillitis*. Although it is observed along with phlegmon of the soft palate and tonsils, it may also occur alone, and may be confounded with the diseases just mentioned. In pure cases the tonsil is only slightly reddened and indistinctly enlarged, while the pillars of the fauces and surrounding parts show the above-mentioned changes. The infiltration of the anterior pillar is generally most intense, as is also the pharyngo-maxillary interstitial tissue. By the inflammatory œdema of the faucial pillar, which often becomes a centimetre thick, and also by the pressure of the accumulating pus, the tonsil may be arched forwards or inwards, or, by the swelling of the anterior pillar, may be completely covered. Peritonsillitis goes on, as a rule, to suppuration, although complete resolution is not impossible. It is sometimes very difficult to determine if pus is present, and where it is. If it be behind the anterior pillar, fluctuation may generally be felt; sometimes it is possible to force the pus backwards, by pressure on the anterior pillar, whereupon it exudes in drops on to the convexity of the tonsil immediately in front of the posterior pillar.

The course of phlegmonous angina is generally acute; it lasts from five to fourteen days. If the affection attacks first the one side and then the other, or if, as is not at all uncommon, a relapse occurs, then several weeks may elapse before it is cured.

The diagnosis depends upon the objective condition, and is sometimes, though only at first, rather difficult. On the other hand, it is generally very difficult to find out where the pus is collecting, especially when the affected parts exhibit uniformly diffuse swelling. The collection of pus can be soonest discovered by palpation, when done in the way recommended by Stoerk. One hand is laid externally under the angle of the lower jaw, the skin and cellular tissue are pressed inwards, in order to act as a support, the point of the index finger being passed from above downwards on to the swollen part, while the index finger of the other hand is laid on the soft palate or tonsil. In this way, when the fingers pressing against one another come upon a soft spot, it is usually the seat of the collection of pus. By using a probe with a broad head, one often succeeds just as well.

The prognosis is nearly always favourable. Still it often happens, when resorption is incomplete, that hypertrophy of the affected organs, especially of the tonsils, remains. In rare cases the affection proceeds to the outside and gives rise to abcesses behind the sterno-mastoid muscle. An unfavourable event, happily of rare occurrence, is choking from obstruction of the isthmus faucium, or from the aspiration of pus. When the larynx participates, asphyxia may occur in a few hours, or even minutes, as happened in a case mentioned by Rühle, though, in my experience, œdema of the larynx is a very rare complication of the disease, and I have not seen a single case in the course of ten years. Fatal cases may also occur from erosion of the carotid, in consequence of pyaemia, gangrene, or of pus descending into the thorax. Why the danger of erosion of the carotid is mentioned as a possible result of peritonsillar abscesses is explained by the anatomy of the parts.

Treatment.—As pharyngeal phlegmon is a very painful affection, an attempt with the previously mentioned abortive remedies is indicated. Ice proves itself to be comparatively the most effectual remedy for averting and mitigating the pain, provided the patients do not come too late under treatment. The uninterrupted wearing of an ice collar, or the application of ice-cold compresses changed every two or three minutes, as well as the sucking of pieces of ice, gargling with iced water or with solutions of alum or tannin (grs. 15 ad ʒi), tend most to mitigate the pain. If treatment does not succeed in preventing suppuration, the appearance of fluctuation must not be waited for, but the part must be boldly incised with a knife, even at the risk of not evacuating pus. I can only agree with Stoerk and Fränkel, when they say, that there is nothing which relieves patients so much as incision, which should be performed at the most prominent spot. If no pus escapes, a second incision must be made at another spot, and a third if necessary; it acts at once by relieving tension of the part, while the blood-letting mitigates, sometimes even cuts short the disease. I have never known harm come from early incision, but have often seen it shorten by several days the dreadful torments of the patients; certainly, it has often happened to me, that unreasonable patients, when no pus appears with the first incision, would not hear of further attempts being made. To accelerate suppuration, poultices are recommended, also warm gargles and inhalations; when fluctuation appears, an incision should be made at once, and as much of the pus as possible allowed to escape by introducing a director, and repeating the process for several days. If the patient refuse to have the abscess opened, he will have to wait for at least eight to fourteen days, until it breaks of its own accord. If respiration is insufficient from closure of the isthmus faucium, tonsillotomy is indicated; if the larynx is also affected, tracheotomy, provided no alleviation is given by scarification of the œdematous parts.

RETRO-PHARYNGEAL ABSCESS.

Synonym.—Post-pharyngeal abscess.

Of the inflammatory affections occurring in the connective tissue in front of the vertebral column, retro-pharyngeal abscess is the most common.

The affection may occur idiopathically, or be symptomatic.

Idiopathic retro-pharyngeal abscess is a disease of childhood. Children six months old, or even younger, are particularly often affected. According to the investigations of Bokai, Kormann, Schmitz, and Pauly, it occurs in consequence of lymphadenitis. According to Henle, there are two lymphatic glands at the level of the second and third cervical vertebræ at each side of the middle line, the volume of which is greatest in childhood; while, after the fifth year only one, or, it may be, neither can be found. These glands must be looked upon as the starting-point of the suppuration.

In the majority of cases, acute catarrhal and phlegmonous inflammations of the oral and pharyngeal cavities, and also of the nose, are considered as the cause of idiopathic abscesses, or lymphadenitis. Abscesses have been repeatedly observed in the course of otitis media and of acute infectious diseases, particularly scarlet fever and diphtheria. Badly nourished scrofulous children, or those affected with eczema of the hairy scalp and face, are more liable to be attacked.

Symptomatic retro-pharyngeal abscess, and gravitation abscess, occur in consequence of diseases of the spinal column, chiefly from periostitis, ostitis, and caries, in scrofulous, rachitic, tubercular, syphilitic, or traumatic dyscrasiæ.

The symptoms caused by retro-pharyngeal abscess may suddenly develop in a few days, or may take weeks and months to appear. A rapid development is observed if the abscess arise from a phlegmon of the connective tissue; in lymphadenitis and gravitation abscess it is slower.

The first symptom is always a difficulty in swallowing. When the development is acute, violent pain is experienced in swallowing, along with febrile symptoms and rapid pulse. The pain is due partly to inflammation, partly to arching forward and narrowing of the pharynx. In chronic cases there may be no, or only slight pain, while the difficulty in swallowing may daily become more apparent from inability to take solid nourishment and by regurgitation of the food taken. Children at the breast cease sucking after several attempts, cry, cough, and vomit up the milk just received. As the abscess enlarges, increasing difficulties of breathing occur even to cyanosis, and by compression of the larynx or its nerves, hoarseness and whistling respiration. The cough, which is also present, is caused partly by the copiously produced mucus flowing down, partly by the larynx participating in the disease, and especially when it is affected with inflammatory œdema. Respiration is often stertorous even when the larynx is unaffected, and is accompanied by a snoring, rustling sound, which is at once increased if the patient bend his head forwards. If the abscess develops near the atlas, the same symptoms are presented as in hypertrophy of the pharyngeal tonsil. As a matter of course, nutrition suffers, especially in children, on account of the difficulty of swallowing; and emaciation, weakness, and anæmia occur when the development of the affection is slow. As a rare complication, paralysis of the facial nerve may be mentioned, being caused, according to Bokai, by inflammatory swelling of the tissues lying round the point of exit of the nerve, or more rarely by inflammation of the nerve itself.

In secondary retro-pharyngeal abscess, symptoms of spondylitis accompany the above mentioned disturbances; these are, inability to move the head to the sides, stiffness of the neck, dislocation and deformity of the cervical vertebrae. According to Neureutter, the snoring which occurs during sleep is exactly similar to that so characteristic in abscesses caused by diseases of the vertebral column.

The objective examination, which is rendered very difficult by the youth of the patients, reveals on the posterior wall of the pharynx greater or less prominences, varying from the size of a pea to that of a walnut, covered with inflamed or normal mucous membrane, according to the stage of the disease. The larger the abscesses the more distinct is the arching forward of the soft palate and tonsils, and the narrower appears the lumen of the inferior section of the pharynx. By palpation, which in deep-seated abscess, and with young children, is the only sure help to diagnosis, the tumour is felt to be elastic and fluctuating. If palpation is impossible, on account of the too great narrowing of the isthmus faucium, or on account of the difficulty of opening the mouth, still, according to Kohts, inspection and palpation of the lateral parts of the neck may aid in determining the diagnosis. The neck shows, even at the beginning of the disease, considerable stiffness, which increases later on, while in the neighbourhood of the lower jaw a distinct swelling arises, which can both be seen and felt. Deep down, moderate swelling of the lymphatics is felt, which is rather resistent and painful; fluctuation, too, may be discovered. By pressing strongly on these glands symptoms of suffocation and asphyxia are elicited, especially if a large abscess be present. There is no doubt that, with older children and adults, examination with the laryngoscope is the best way to ascertain the situation, size, and consistence of the abscess.

The diagnosis of retro-pharyngeal abscess is not, as a rule, difficult, if attention be paid to the subjective symptoms, and the results of objective examination. The diagnosis, however, offers special difficulties, if the symptoms are concealed by other affections of the mouth and throat, particularly scarlet fever and diphtheria, or if the larynx is affected at the same time. The greatest importance must always be attached to digital examination, which must eventually be made, the patient being anæsthetised, and to the presence of fluctuation, which is observed in no other pharyngeal affection except pharyngeal hæmatoma, which will be spoken of further on. Secondary gravitation abscesses offer, as a rule, no diagnostic difficulties as regards the simultaneously existing symptoms of spondylitis. It is quite possible to confound these abscesses with soft tumours, especially with retro-pharyngeal lipoma, once observed by Taylor, but puncturing will make the diagnosis certain.

The prognosis of idiopathic retro-pharyngeal abscess is pretty favourable, when development of the symptoms is slow and operation energetic. The more tender the age, and the more difficult the access to the tumour, the more unfavourable are the prospects, and the more violently and more acutely do fatal symptoms develop. The fact that in secondary abscesses there is an absolutely bad prognosis, is due to the nature of the diseases which cause them. Deeply seated secondary abscesses, as well as evacuation into the larynx and trachea, may lead to a fatal issue from asphyxia, or from traumatic pneumonia.

Treatment.—Treatment must be directed to preventing the development of the abscess; unfortunately the attempt to stop the progress of lymphadenitis by ice succeeds but rarely, owing to the youth of the patients. Schmitz and Gautier propose painting the soft palate and posterior wall of the pharynx with tincture of iodine, or iodide of potash solutions, to absorb and more rapidly mature the abscesses. As soon as pus has formed, the abscess must be opened. The opening may be made either with a knife, which is covered up to the point with plaster, with a concealed knife, which is pushed forward from its sheath at the moment the operation is performed, or by means of a trocar. In any case the instrument must be inserted not more than three or four millimetres from the middle line, in order to avoid the internal carotid. In order that the abscess may not turn aside when the incision is being made, it is proposed that an assistant be directed to make pressure from the outside in the neighbourhood of the angle of the lower jaw, and so steady it; while, to prevent the pus being inspired, the patient's head should be bent forwards immediately after the incision is made. It is better, however, that the operation be done with the head already bent forward. Others, and amongst them Schmitz, advise pressure of the finger on the epiglottis, to prevent the pus entering the lungs. If, in spite of every caution, pus is drawn in, then, according to Bokai, an induced current must be employed, by means of which threatened asphyxia has been repeatedly and successfully averted.

Whether merely a puncture or a free incision is indicated, depends on the size and maturity of the abscess, and should it refill, a second operation is necessary. Gravitation abscesses should be opened only when symptoms appear which threaten the life of the patient.

The after treatment consists in syringing the throat with warm water, or with a 4 per cent. solution of boracic acid, while for older children the use of gargles or inhalations is indicated.

EXUDATIVE PHARYNGITIS.

Synonym.—Pharyngitis Exudativa.

On the mucous membrane of the pharynx, just as on that of the mouth, there occur small vesicles, vesicular eruptions, and other exudations.

Miliary eruptions occur chiefly on the soft palate, tonsils, and posterior wall, and are seen as small projections of the size of millet seeds. They are filled with a shining transparent material, and give rise to few or no symptoms of sore throat, become dry, and disappear, leaving behind no loss of substance or ulcer.

Of more frequent occurrence is herpes (*angina herpetica, angine conenneuse*). It appears most usually along with herpes labialis, but is, however, also met with as an independent disease.

There is still great diversity of opinion as to its etiology. While some attribute its occurrence to chills, damp air, or cold wet weather, others connect it with over exertion, mental affections, or uterine disorders. In support of the influence of the weather, Semon's observations may be mentioned, that pharyngeal herpes occurs in England, mainly in spring and autumn. Herzog believes that it is excited by irritable conditions in the region of the nerve branch supplying the pharynx, and of the posterior nasal nerve, as well as of the Vidian nerve, and supports his opinion that the disease is neuropathic, on the fact that it is unilateral, that it begins with violent unilateral headache, and also on the circumstance, that after its disappearance, paralyses of the soft palate remain.

The disease begins, as a rule, suddenly, and is generally accompanied by feverish symptoms and headache, the latter being often unilateral, and having the character of megrim. Soon there occur burning pains in the neck, which are considerably increased by deglutition, increased secretion of saliva, and hawking.

On examination small vesicles are seen, occurring principally on the isthmus and pillars of the fauces, the tonsils, and uvula, varying in size from that of a pin's head to that of a lentil seed. Yellowish in colour, arranged in groups, and surrounded by a red inflammatory halo, they vary very much in number, sometimes only a few, sometimes very many being present. The eruption is usually unilateral, but may occur on both sides; it does not come out all at once, but generally for two or three days further outbreaks take place. The tonsils are enlarged and inflamed, after a short time the vesicles burst, leaving at first red spots which soon become covered with a white opalescent mass, and form circular ulcers, which, when several vesicles coalesce, may easily be thought to be diphtheritic. The pain at this stage is extremely great. In from eight to ten days, healing takes place, after other herpetic eruptions have appeared on the nose and lips, or on the prepuce, or at the entrance to the larynx. According to Herzog, the herpetic ulcers may penetrate to a considerable depth, and cause perforation of the soft palate, while Hallopean states that gangrene may set in and cause death. Although, as yet, I have never met with such results, I have but little reason to doubt them, as I myself once lost a patient—an oldish lady—from pyæmia, in consequence of a unilateral herpes of the face.

The course of the disease is generally acute, still it may run a chronic course, and follow a certain type.

The diagnosis presents no difficulties, if one discovers vesicles in the throat or mouth, near the already formed ulcers; but if this be not the case, then a certain diagnosis is impossible, and confusion with syphilis or diphtheria, is at least excusable.

Treatment.—The treatment must be local as well as general. In cases where the disease can be traced to excessive chills and rheumatic influences, salicylate of soda is recommended, but more specially tincture of aconite. When neuropathic in character, good results have been got from arsenic. The local treatment is limited to gargles of easily dissolved mucilaginous remedies, such as biborate of soda grs. 5 to decoction of althea ℥i ; since other remedies, as a rule, rather increase than allay the pain.

Larger vesicles occur also in the throat in erysipelas pharyngis, which was mentioned with phlegmon on account of the predominating excessive diffuse inflammation of the mucous membrane; they occur, however, chiefly in *pemphigus*. The affection appears partly as the forerunner of pemphigus cutaneus, and partly as an independent form.

The etiology is still very obscure; there is no doubt that it may occur in a syphilitic dyscrasia, and also sometimes in consequence of uterine disorders.

Like herpes, it generally begins with fever, rigor, nausea, and vomiting. On the mucous membrane of the soft palate and uvula there appear vesicles, varying in size from that of a pea to that of a nut, containing a milky white material, whose epithelial covering, after a short time, becomes dim, then macerates and is shed off, leaving behind sharply defined red erosions, covered with a white croupous or aphthous-like membrane, which either heal rapidly or become converted into ulcers. It is easily understood that pemphigus of the throat may cause not only pain on swallowing and indistinct speech, but also by encroaching on the space of the pharynx render deglutition difficult, and under certain circumstances give rise to dyspnœa.

Pus filled vesicles—*pustules*—arise, either in consequence of a change taking place in the contents of the eruptions before mentioned, or from variola.

Pharyngeal variola proper, which must be distinguished from the inflammation of the mucous membrane accompanying small pox, is almost always a co-symptom of variola of the skin, the general infection being, in rare cases, localised only or chiefly in the pharynx, as was observed by the author in his own case. After rigor, pains in the loins and fever (as high as 106° F.), there develop on the reddened marbled mucous membrane of the soft palate and posterior wall, and often also of the mouth, round nodes about the size of a lentil seed, which generally on the second or third day exhibit a vesicle in the centre, and become changed into pustules. Round the pustules is a bright red halo, the salivary and lymph glands swell up, there is excessive salivation and violent pain on swallowing, which is still more increased when the pustules burst. When the eruption is profuse on the posterior wall, danger of choking may arise from œdema of the larynx.

In the naso-pharynx, according to Wendt, the affection goes on to hyperæmia, and hæmorrhage and diffuse swelling of the mucous membrane occur, with the subsequent production of a layer of false membrane and ulceration. These ulcerations, like the ulcers from real pustules, may penetrate deeply and cause perforation of the soft palate.

The diagnosis is easy, when variola of the skin is also present, but without it, only possible by taking into consideration the febrile commencement, the prevalence of a small-pox epidemic, and the results of objective examination.

Treatment.—The general treatment consists in overcoming the fever, and allaying the extreme pain in the head and neck: the local treatment in diligent cleansing of the pharynx with unirritating antiseptic gargles.

Aphtha is to be considered as an inflammatory process, with deposit of, as is affirmed, a croupo-fibrinous exudation in the upper layers of the mucous membrane.

As regards the etiology, what has been previously said must be referred to. The favourite situation of aphtha is the soft palate and the tonsils, and less frequently the posterior wall, from which it sometimes extends to the larynx. Sometimes aphtha runs an acute course, sometimes chronic, with relapses at long intervals. The symptoms consist in burning, darting pains on swallowing and speaking, and the necessity for hawking.

The diagnosis is not difficult in the acute form. In the chronic it is often possible only after continued observation. Aphthæ are almost always confounded with syphilitic patches, and hence patients are subjected to wholly useless antisyphilitic treatment. I was once successful in immediately and permanently removing the disease in a patient, in whom the aphthous process was localised to both tonsils, and who was very much reduced by repeated inunction cures, by amputating the tonsils. As for the rest, the treatment is the same as in aphtha of the mouth.

Another exudative affection is *pharyngitis fibrinosa*.

By this is understood neither the earlier "pharyngeal croup" nor the diphtheria of to-day, but an affection which is not infectious, and which depends upon an increase of the inflammation. One finds delicate hoar frost-like deposits, or circumscribed croupous exudations, often in consequence of catarrhal or phlegmonous angina; I repeatedly saw and examined them along with follicular ulceration of the tonsils. The occurrence of a non-infectious croupous angina is most decidedly affirmed by competent authorities, particularly Bamberger and Fränkel. Its seat is the tonsils and posterior wall, but the process may also spread beyond this. An artificial croupous exudation is always produced by galvano-caustic operations.

The membranes which form can easily be stripped off, or they are shed of their own accord after a short time, and sometimes become renewed. The subjective symptoms are the same as in all acute anginæ.

The differential diagnosis can scarcely be made on account of the extraordinary extension of diphtheria. According to my own observation, it is only possible when follicular ulcerations are also present, since along with them real diphtheria never occurs. Fränkel says also that the local condition is of no value in making the diagnosis, which can generally only be made at a later period.

The treatment is the same as in acute pharyngitis.

DIPHTHERITIC PHARYNGITIS.

Synonym.—Pharyngitis Diphtheritica, &c.

By the names *Pharyngitis* or *Angina diphtheritica, pharyngeal diphtheria, diphtheria,* and *cynanche contagiosa,* is meant an infectious disease, which is characterised by the appearance of pseudo-membranes on the mucous membrane of the pharynx.

Diphtheria is generally a primary independent disease. Secondarily it accompanies scarlet fever, typhoid, and other severe infectious diseases, or is associated with wounds of the throat. It is originally a local infectious disease, which spreads from the parts first attacked over the whole body, thus becoming a general disease. It belongs to the class of miasmatic-contagious diseases, and spreads by the intercourse of healthy persons with persons and objects infected with diphtheria, by means of the secretions and breath of the patients, and by their clothes. The diphtheric poison can retain its power of infection for months outside the human body, as is proved by the outbreak of the disease after frequenting an infected room, and that men may become infected by diphtheritic animals, especially by hens, has been observed by Gerhardt. Incubation takes, as a rule, from two to seven days, but it is affirmed it may last fourteen days. When the transference is very direct, as, for example, with physicians, the disease generally begins on the third day, provided that predisposition to it exists.

Diphtheria most usually occurs as an epidemic; it is sometimes localised at first to one storey, or to one house, and extends from thence throughout a street, a town, or even a whole province. Individual epidemics show great differences in regard to the severity and malignity of the disease. In the large towns on the continent it prevails year after year. It is most frequent during the colder part of the year, and commonly attacks children. While infants and children under one year are seldom attacked, the greatest mortality occurs between the ages of one and five: next follows the period of school life from five to ten, later years show a smaller number of deaths, while other sections are on the whole pretty equally represented. Constitution has no influence, but there is no doubt that children who are affected with acute inflammations of the pharynx and the nose, or with bronchitis, present a more favourable soil for the infectious material. According to Monti, sudden chills, excessive dampness of the atmosphere, rapid changes of temperature, defective hygienic conditions, damp, musty, or newly built dwellings, dirt, badly cleaned sewers and canals, are favourable to the development of diphtheria, while unfavourable to it are pure dry air rich in oxygen, and good sanitary conditions. A second attack in the same patient occurs very rarely, and I must, on the ground of personal experience, confute as absolutely incorrect the opposite assertion, that one attack predisposes the patient to another.

Latterly the number of those who consider that diphtheria originates sporadically, *i.e.*, without previous infection, has greatly increased. According to Heubner, it is not at all impossible that spasm of the superficial capillaries of the pharynx, excited reflexly by cold, may be followed by complete cessation of the circulation and croupous exudation.

There is no question that has occupied the medical world more in late years than that as to the etiology of diphtheria.

Without entering on the numberless more or less ingenious hypotheses, as, for example, the generation of diphtheria from potatoes or from decayed vegetables, bread or milk, let us turn to the main question, whether or not diphtheria is really excited by a specific microorganism—the *micrococcus diphtheriæ*. Heubner declares himself to be directly opposed to this theory, and Löffler, who again lately tried to solve the problem, has been unable to come to any certain conclusion.

Quite recently Emmerich has found the true diphtheria bacteria, has produced cultivations, and inoculated them with positive results. In the blood they may not be found at all, on account of their exceedingly small number, but they occur in great numbers in the kidneys. The diphtheria fungus may have its seat of germination outside the human body, in the floors, and especially in the ceilings of dwelling-houses. The greatest mortality occurs, therefore, during those months when people are compelled to occupy heated rooms; at seasons when rooms are not heated, the contagion loses its malignity and danger.

Whether Emmerich's investigations have more demonstrative power than those of his predecessors, must be learned in the future; till then, it must remain a doubtful point, whether the micro-organisms are not rather the products than the causes of diphtheria.

However little we as yet know about the poison of this disease, we have certainly been very much helped by Heubner's classical experiments as to the nature of diphtheria. There can be no doubt, according to him, that the diphtheritic affection of the mucous membrane, is an intimate combination of inflammation and necrosis. The necrosis is the result of local disturbances in the circulation, going on to complete stasis, with co-existent disease of the vessels themselves. Hence, it leads not only to exudation of white blood corpuscles into the mucous membrane, but also to the existence of a croupous pseudo-membrane on the mucous membrane itself. The process, when localised to the superficial layers, is then known as genuine primary superficial diphtheria; when the tissue of the mucous membrane or the parenchyma of the affected organ decays and becomes infiltrated with a fibrinous exudation, then the affection is called real tissue diphtheria. What really gives rise to relapses is the persisting diseased condition of the walls of the vessels. As soon as they recover from disease and become normal, the development of the croupous membrane also stops. In *tissue-diphtheria* proper, the vessels are altogether closed by the necrosed tissue, and here the death of the tissue, which in croup does not affect the underlying diseased vessels, causes also the final death of the closed up vessels. Wherever tissue-diphtheria extends, everything must perish; if the shedding is completed, loss of substance remains, which becomes an ulcer, while limited suppuration takes place in the surrounding parts.

From anatomical and clinical reasons, it is best to divide diphtheria into two forms—*croupous* and *septic*; the presence of a pseudo-membrane being characteristic of both.

Although it cannot be denied, that sometimes infection with diphtheritic contagion does not cause pseudo-membranous, but rather catarrhal or phlegmonous inflammation of the mucous membranes, still it is not altogether advisable that such diseases be also designated *diphtheria*, as thereby great scope would be given to personal arbitration.

Diphtheria of the pharynx in children sometimes begins suddenly, with rigor, general discomfort, heat, thirst, vomiting, headache, and pain in the neck, back, and lumbar region; in older children and adults it begins more gradually, with slight fever, just as in catarrhal angina. The outbreak of diphtheria is not infrequently accompanied by peculiar appearances on the skin; most often there is an erythematous redness on the arms, neck, and throat, or on the breast and groin, so that one may be in doubt whether it is not a case of scarlet fever. Urticaria-like efflorescences on the extensor surfaces of the extremities may also be observed. The appearance of petechiæ and blood extravasations on the trunk or on the extremities must be considered as a more certain sign of the severity of the case, and of its generally septic course. In a short time, generally in a few hours, after the commencement of the general symptoms, the local manifestations become prominent. These consist in burning and dryness in the throat, with pain on swallowing, or in the feeling of a foreign body in the pharynx.

Objective examination, if the case is seen very early, may reveal the appearances of acute catarrh only, but in most cases distinct signs of exudation are already present. These may be limited to individual parts of the pharynx, but may also be diffuse from the beginning. In mild cases the exudation is localised to one or both tonsils, to the uvula, the pillars of the fauces, or the posterior wall. The tonsils appear red and swollen, and adhering to them are one or several disseminated, milky white patches, or a hoar frost like coating. If the uvula is attacked, it appears œdematous, and altogether or only partially enveloped by a white mass, which sends out branches to the pillars of the fauces and central part of the soft palate. On the oral division of the reddened posterior wall, punctiform, or more diffuse exudations, are very distinctly seen. In severe cases both tonsils, or the whole of the soft palate and the posterior wall, are attacked at the very beginning. The pseudo-membranes have their seat, as a rule, on the epithelial layers, but adhere pretty firmly to the underlying parts, and very often penetrate into the deeper layers of the mucous membrane, and the parenchyma of the organ affected.

Saliva and mucus are copiously secreted, and the posterior wall particularly appears invested with a thick green purulent secretion. The taking of food, on account of considerable swelling of the affected parts, and paralysis of the soft palate, is not only painful, but also rendered most difficult, especially if the mouth and mucous membrane of the hard palate and of the cheeks or lips are also affected with diphtheria.

A symptom which is always present, is swelling of the lymphatic glands. It occurs in even the mildest cases, while in severe ones it attains a most extraordinary intensity. The glands below the angle of the lower jaw, the submaxillary gland, and the superficial and deep glands of the neck, swell up, are very painful, and form large hard prominences or bead-like masses, which sometimes occupy the whole space between the lower jaw and the clavicle. They either remain hard during the whole course of the disease, and become resolved as the patients recover, or become soft, and evacuate sanious pus mixed with shreds of dead cellular tissue.

The fever is sometimes insignificant, but may, as the exudation progresses, rise from 103° F. to 106° F., or is very considerable from the beginning. In general, the fever at the beginning of the disease is in distinct proportion to the quantity of the diphtheritic products, and rises or falls as the exudation increases or diminishes. If the exudation comes to a stand still, although only temporarily, the fever often falls of its own accord to below normal. Regular typical curves which occur in other febrile diseases, such as pneumonia and typhoid, are not observed in diphtheria. Sudden fall of temperature is caused by commencing paralysis of the heart, or by poisoning from carbonic acid, in consequence of diphtheria of the air passages.

In like manner is the pulse affected. At first it is usually hard, as in any febrile disease. In older children it increases to 100 or 120, in younger children to 130 or more; in adults, 90 to 110. As the fever diminishes and the membranes are shed off, it becomes soft and less frequent. In the septic form, the pulse at first is very rapid and small, even when the temperature is considerably increased. If the heart fails, or if collapse or carbonic acid poisoning come on, the pulse becomes extremely frequent, 130 to 180, is intermittent, and scarcely to be felt.

Speech and respiration are subject to many changes.

Speech is impeded, anginous and nasal; when the nose is also affected, it is excessively nasal and indistinct. Respiration is snoring and rustling, according to the degree of swelling of the pharynx. Dyspnœa occurs generally when there is complete closure of the isthmus, and when the larynx or the trachea and bronchi are also affected.

The digestive organs are more or less affected. The vomiting, which occurs early in the disease, is a febrile symptom, and is of central origin; later on, it is often of a reflex nature, excited by the irritation of incompletely shed membranes flapping about, or occurs as the first symptom of commencing uræmia or septicæmia. The appetite is usually completely lost, so much so, that patients must be forced to eat, thirst is increased, and the bowels constipated, though in some epidemics diarrhœa occurs, and is then generally a sign of collapse or commencing blood poisoning.

The urine is considerably diminished, is thick and concentrated, rich in uric acid, or in phosphates, and sometimes from the earliest stage contains albumen.

As the disease continues, almost every case is different from the last.

In the mildest cases, the affection begins to resolve after from three to six days, the membranes are shed, and the patients recover. In moderately severe cases the exudation spreads from the place where it originated to the surrounding parts, or the membranes become reproduced, with renewed rise of temperature and increased lymphatic swelling; there is a fetid smell from the mouth, and the general condition gives rise to anxious fears. In spite of this, recovery may take place, if no further exudation or complication steps in after shedding of the thick layers. In very severe cases the membranes become reproduced for the third, fourth, or even fifth time; the disease is protracted for weeks, and may, in consequence of long-continued rise of temperature and loss of appetite, endanger life from debility or from the larynx becoming affected; nevertheless, in such cases recovery may still take place, even although secondary diseases of all sorts occur.

Even in the mildest forms an extension generally takes place towards the superficial and deep parts; in severe cases this is also the case. This enlargement is the consequence of several small neighbouring exudations becoming apposed, or of the participation of new parts of the mucous membrane. The membranes attain a very considerable thickness in the later stages, by absorption of water, and often form layers as thick as the finger, or sloughs, like oyster shells, projecting considerably above the level of the mucous membrane, which are partly separated from the layers underneath, and partly remaining in connection with them. At the same time their colour changes; the membranes, which at first were white or yellowish white, become, from an admixture of blood, brown or brownish black, discoloured, and bad smelling. On the inflamed parts of the pharynx, which are free from exudation, greater or less extravasations of blood are seen.

The croupous exudation, which, as was previously affirmed, spreads only in the pharynx, may, however, also extend to other organs.

Diphtheritic disease of the nose, is one of the most frequent complications. It spreads to the nasal fossæ, either from the posterior wall of the pharynx, or from the posterior surface of the soft palate, and attacks first the posterior parts of the nose; later on the anterior parts also become affected. Diphtheria of the nose very seldom occurs primarily.

The fact that the nose is attacked, is indicated by increase of fever, and symptoms of acute catarrh; often the first symptom is epistaxis, in consequence of the hyperæmia, which occurs before the croupous exudation. If the exudation has occurred, respiration through the nose is rendered very difficult, or altogether impossible; from the nostrils issues a watery mucus, sometimes glutinous secretion, in which, after two to four or more days, distinct pseudo-membranous shreds are seen, and which causes excoriation of the nostrils and lips. When the tubes are affected, tinnitus, earache, and deafness appear, and in very severe cases, purulent middle ear inflammation, with perforation of the drumhead. If the diphtheritic inflammation invades the nasal duct, diphtheria of the conjunctiva and cornea may also set in. As decomposition of the membranes goes on, the secretion becomes fetid, ichorous, and bloody, the skin of the nose swells up, becomes red, and hæmorrhages become more frequent and more severe.

Examination reveals considerable narrowing of all the nasal passages, by pseudo membranous collections, on the turbinated bones and septum; the membranes are sometimes diffuse, sometimes circumscribed, and are shed in adherent shreds, or as masses of detritus; their reproduction occurs in the same way as in the pharynx. Nasal diphtheria may be recovered from, or may give rise to septic infection, and it may also occasion erysipelas of the face and meningitis, with fatal results.

Far worse than that of the nose, is diphtheria of the larynx, trachea, and bronchi. The worst cases of pharyngeal diphtheria are not always the ones which result in these most dangerous combinations; on the contrary, they come under observation just as often in cases of thin hoar frost like and circumscribed exudations, though, in general, the possibility of the disease spreading to the larynx, is to be feared in all forms. The affection of the larynx usually sets in from the third to the eighth day, but may begin earlier or later. Extension downwards is what usually occurs, but the process may begin in the trachea and larynx and spread upwards, as I have myself repeatedly seen. The disease of the larynx sometimes begins with vomiting and high fever. Hoarseness is a constant symptom (catarrhal stage). It is in the beginning often so insignificant, that it is only recognised by the experienced ear of the physician; in a few hours, or perhaps days, it more and more assumes the character of aphonia. At the same time a very barking cough, with scanty expectoration and increased difficulty of swallowing, sets in. Laryngoscopic examination, which unfortunately with young children is often impossible, reveals in older ones, or in adults, besides diffuse redness and swelling of the mucous membrane, circumscribed aggregations, varying in size from that of a pea to that of a bean, or membranes adhering to one another, covering the vocal cords, epiglottis, and ventricular bands, and sometimes the whole larynx. I have repeatedly seen aggregations in the interarytenoid region, on the aryepiglottic ligaments and epiglottis, that did not spread as the disease went on, nor give rise to dyspnœa. The amount of obstruction to respiration depends on the thickness and extent of the exudation, and also very particularly on the age of the patient. In children, obstruction of respiration very soon makes itself evident. Breathing is noisy, whistling and stertorous, the lower parts of the thorax sink in, the larynx is deeply drawn down towards the jugular fossæ with each inspiration, while all the accessory muscles of respiration are called into play. Children are very restless, turn themselves from one side to the other, sit up in bed, throw back the head, catch at their throat, demand to be taken into their mother's arms, and then back to bed again (stenotic stage). As in real croup so also in diphtheria, dyspnœa comes on in paroxysms, recurring at longer or shorter intervals, till gradually the demand for breath becomes continual, and the disease enters its last stage, that of asphyxia.

The indescribable anxiety and restlessness gradually give place to a condition of repose, the hitherto reddened countenance becomes pale, the lips, fingers, ears, and lower extremities become cyanotic, the body is covered with a cold clammy sweat, the temperature falls, cough becomes less frequent, or stops altogether, carbonic acid poisoning gradually increases, and from apathy, somnolence and coma develop, and in one or two days, often even in twelve hours, the patient dies.

On account of the width of the adult larynx, marked dyspnœa seldom occurs in circumscribed exudation; but, of course, when the disease is diffuse, and the pseudo-membranes are pretty thick, it is never absent.

The trachea is not, as a rule, attacked, until it has been operated on for diphtheria of the larynx. If the trachea is attacked earlier than the larynx, as I have often seen, the movements of the larynx and of the trachea downwards are far more slight than when the obstruction is in the larynx itself. Diphtheria of the trachea, after tracheotomy has been performed, makes itself known by rigors and increased rise of temperature. The secretion coming through the tube becomes scanty, and is mixed with croupous shreds, then stops altogether. When the bronchi are affected, branching casts are sometimes expectorated; in tracheitis diphtheritica, thick tube like cylinders. According to Monti, the lungs in front of the manubrium sterni, and in the first intercostal space, give a dull, and at the same time tympanitic note, the breath sounds are here and there absent, or are replaced by rustling moist sounds.

Among the subjective symptoms, the appearance of new difficulties of respiration, in spite of tracheotomy, is characteristic. They are generally observed from the second to the fourth day, or even later; respiration, which at first after operation was normal, becomes frequent, superficial, and laboured; all accessory muscles of inspiration are brought into play, the ribs and diaphragm are drawn in, the want of breath increases, jactitation and restlessness reach their highest pitch, gradually a corpse-like pallor comes on, the children fall, sometimes after repeated paroxysms of suffocation, into somnolence and coma, and die from carbonic acid poisoning.

The diphtheritic exudation sometimes attacks also the sexual organs and the external skin.

In some, particularly in septic cases, the disease attacks these parts primarily. Females are more frequently affected in this way. The labia minora swell up, the mucous membrane is reddened, and emits a purulent secretion. Burning and straining pains, difficulty of micturition, strangury and spasm of the bladder are experienced, and on the inner surface of the labia majora and minora, whitish yellow aggregations are formed, extending over a large surface, or scattered about, which are shed after a time, are renewed again, and then heal, or may give rise to sepsis and gangrene. The diphtheritic process often invades the anus and neighbouring skin, and produces erosions and exudations there. Swelling of the inguinal glands is hardly ever absent.

In males diphtheria confines itself principally to fissures and injuries of the prepuce. The prepuce, and also the skin of the penis, becomes swollen and œdematous, and cannot be drawn back, and between the prepuce and the glans a purulent fluid appears. The internal surface of the prepuce is covered with a pseudo-membrane, the inguinal glands swell up, and there is violent pain on micturition. By breaking down of these membranes, and the underlying mucosa, considerable loss of substance may be produced.

Diphtheria of the skin must be looked upon as a very rare form of the disease.

It has been affirmed by several authors, chiefly Fr. Seitz, that diphtheria may develop on perfectly healthy skin. According to Seitz, red, round, or elliptical, irregularly formed spots appear, with a yellowish white prominence in the centre, like an urticaria wheal. On this spot a yellowish brown fibrinous exudation forms, which after drying up is shed, and leaves an ulcer in its place. After the ulcer heals, a brown pigment spot remains visible for a long time. The skin, where excoriated and devoid of epithelium, is more often attacked, both primarily and secondarily, than where it is intact.

Eczematous patches behind the ear, or on the ear itself, at the corner of the mouth, the entrance to the nose, and chafed spots in the inguinal, crural, and perineal regions, are favourite points of attack ; also on leech bites, operation wounds, or places where blisters have been, the infection sometimes shows itself first.

We will now turn to the second form of diphtheria, *the septic.*

It develops, as a rule, out of the croupous form, with or without simultaneous gangrene of the affected parts, or it sets in at the very beginning as primary septic diphtheria.

The development of the septic from the croupous form seldom occurs before the end of the first week ; generally however later, and not for several weeks. It is due to putrefaction and decomposition of the partly shed, partly adherent membranes, which assume a brownish black appearance, are of a pultaceous consistence, and emit a highly penetrating fetor. Should the septic material contained in these putrefying membranes reach the blood and be communicated to the various organs, then diphtheria, from being a local affection, becomes a general disease. The resorption of the septic poison may take place from the nose, throat, larynx, trachea, and bronchi, and also from the skin or the genital organs. It manifests itself usually as adynamia, and rise of temperature, without however, as in the other form of sepsis, going on to distinct rigor. The previously excited and refractory patients become suddenly quieter and more indifferent, the skin becomes pale, the eyes hollow and dull, complaints about swallowing and other symptoms cease, the examination of the throat hitherto so difficult is allowed without resistance, and every kind of nourishment is refused with signs of repulsion. The brain, which at first was clear, gradually gives way, somnolence comes on, sometimes interrupted by vomiting and diarrhœa. Consciousness is sometimes retained till shortly before death, which usually takes place from paralysis of the heart or œdema of the lungs, with steady decrease of the extremely frequent and small pulse, and outbreak of perspiration. The mucous membrane of the part attacked is covered with larger or smaller ecchymoses, otherwise it is but little changed. Here and there the epithelium is wanting, the surfaces shedding slightly, and having superficial erosions, or deeper losses of substance penetrating into the tissue beneath. Although, hitherto, lymphadenitis has kept within bounds, the glands now swell up into large shapeless tumours, which, being very painful, hinder the movements of the head and opening of the mouth.

Quite otherwise is the condition, if the croupous form first passes into the gangrenous and then into the septic. Although the danger of gangrene, in very diffuse exudation and infiltration of the parenchyma, is greater than in localised, there are nevertheless exceptions, and the time at which gangrene begins varies very much. As a rule, it begins when the diphtheria has existed for several days, or even one or two weeks, although in rare cases it may appear much earlier. The author once saw, in the case of a boy of twelve, who had pseudo-membranes scattered over the tonsils, the whole soft palate become gangrenous on the second day. It is exactly in such cases that combination of necrosis with inflammation occurs in a very distinct manner. The occurrence of gangrene proclaims itself first of all by objective symptoms; the tonsils, which up till now were moderately swollen and covered with pseudo-membranes, become very much larger, close up the isthmus, and squeeze in the uvula, which is œdematous, and, like the tonsils, covered with exudation. If the gangrene attacks the soft palate, which is by no means always occupied by thick pseudo-membranes, then there appears on some part of it a bluish black livid spot, distinctly defined from the surrounding surface, over which the mucous membrane is raised in the form of a cyst, with dim discoloured contents. The parts round this cyst also soon become discoloured, and after it gets emptied the whole part affected is changed into a greyish black, soft, pulpy mass, containing many blood extravasations, which bleeds either spontaneously or on being touched. If the patient does not yet succumb to the disease, as sometimes happens, then deep losses of substance and perforations occur from the mortified masses becoming separated; in very rare cases a line of demarcation forms, the gangrenous part separates, and the patient escapes with more or less destruction of the parts. After the gangrene comes on, the fetor from the mouth is unbearable, salivation is profuse and ichorous, and the general condition assumes the already mentioned adynamic character. The gangrene spreads to the nose, from which a sanious discharge comes, excoriating the skin; there are repeated bleedings from the throat and nose, and the patients generally die in the course of from three to eight days, sometimes with metastatic affections in the lungs and spleen.

The course of primary septic diphtheria is very similar, though differing in some points.

Why, in some epidemics, this form is exceedingly frequent, in others altogether absent, we do not know. The disease begins with violent fever and vomiting, several times repeated, accompanied by nausea. The temperature reaches 104° F., or even 106.5° F., the pulse is very frequent, and from the beginning very easily compressible; respiration is quickened. The general condition is very low from the first, the patients are quiet, apathetic, and sleepy. Pain on swallowing is but little complained of.

Examination reveals livid blue grey colouring and ecchymosis of the affected parts, which are covered with fibrinous, greasy, nasty looking exudations. The swelling of the lymphatic glands is also in this form very marked. The further course of the disease is the same as in the other already described septic forms, only the fatal issue as a rule takes place earlier, usually after two or three days.

The symptoms just discussed do not, however, exhaust the morbid appearances of diphtheria. Diphtheria is an extremely malignant disease, which not only during its course gives rise to many complications but also, for many weeks after apparently complete recovery, is followed by a succession of secondary affections.

One of its most frequent accompaniments is disease of the kidneys. This affection manifests itself first of all by albuminuria. The albumen contained in the urine of the first day is to be considered as a symptom of fever, and traceable to hyperæmia of the kidneys; the adherents of the fungus theory consider it due to immigration of micrococci or to a parenchymatous inflammation excited by them. Albuminuria occurs more surely and earlier, the more intense the exudation is: in septic cases it is never absent, so that Unruh maintains that albuminuria is the first and only absolutely certain symptom of general infection having taken place, which therefore is excluded if albuminuria be absent. I must, however, contradict him, since quite slight as well as very severe cases of the croupous form occur having albuminuria, without going on to general infection. The quantity of albumen varies very much, and fluctuates according to the local and general symptoms. Albuminuria very often occurs when the larynx is attacked. According to Monti, it is the consequence of hyperæmic engorgement of the kidney, excited by stenosis, which disappears when the circulation again becomes normal. The more severe forms of albuminuria are caused by parenchymatous nephritis, and by swelling and fatty degeneration of the renal epithelium, and Fische found in all his cases parenchymatous and interstitial nephritis, but no micrococci. The secretion of urine is diminished, even completely suppressed, the amount of albumen is very considerable, the urine contains blood corpuscles, fattily degenerated renal epithelium, and different forms of casts.

Although by very many authors a rather subordinate significance is directed to diseases of the kidneys, yet I must, on personal grounds, hold quite another opinion. I have repeatedly seen children and young people die of uræmia. All these patients showed very extensive local symptoms; in one case not only both tonsils, but also the whole of the soft and hard palates, were covered with layers as thick as the finger In all cases the course was similar. After albuminuria had set in, several days after the beginning of the diphtheritic exudation, the amount of urine always diminished, while the albumen progressively increased; the first symptom of uræmia was always nausea and repeated vomiting, as well as complete loss of appetite; the pulse was sometimes hard and tense, sometimes soft, small, and frequent, soon somnolence and drowsiness set in with continued vomiting, and in one case eclampsia, and death followed with complete loss of consciousness from loss of cardiac power, and it seemed to me remarkable that neither œdematous swelling nor dropsy were ever present.

Complications very often arise in connection with the heart.

During the course of the disease, as well as during convalescence, patients may suddenly die with symptoms of acute paralysis of the heart. As a rule no symptoms, or at least no special symptoms, precede this event. The patients complain suddenly of palpitation, want of breath, oppressive and extreme prostration; they become pale, cyanotic, and cold; the pulse becomes extremely frequent, and can hardly be counted; they become unconscious, and death occurs generally in a few hours. In the cases running a less rapid course, the symptoms of cardiac paralysis set in gradually, and only after several days lead to a fatal issue. These sudden deaths are explained partly on the hypothesis of fatty degeneration of the cardiac muscle or thrombosis of the ventricle, partly of paralysis of the vagus and cardiac ganglia.

But of far more frequent occurrence than acute, are chronic cardiac affections. According to the investigations of Leyden and others, myocarditis seldom begins before the end of the first week. It manifests itself by the pulse becoming small and frequent, reaching from 120 to 180 beats per minute, or becoming intermittent, these symptoms being more evident on movement of the body. The heart's beat is weak, in a little time the heart's dulness increases towards the right, the sounds become impure, and blowing murmurs are heard. Subjective symptoms are absent or are very slight, often there is apathy and disinclination for bodily exertion, and according to Unruh, albuminuria sometimes occurs. The affection may be recovered from after several weeks, or even several months' duration, or may end fatally from dropsy or cardiac paralysis. It is my conviction that many cases of so called myocarditis die, not from degeneration of the cardiac muscle, but from paretic conditions of the cardiac nerves.

Of complications in connection with the respiratory organs, pneumonia is relatively the most frequent. I have observed it most commonly in tracheal and laryngeal diphtheria, and after tracheotomy, in consequence of blood or pieces of membrane being drawn into the lungs. It is as a rule lobular, unilateral or bilateral, and generally occurs in the lowest part of the lungs. Several times I have seen fatal traumatic pneumonia in consequence of bilateral paralysis of the recurrent nerves.

As further complications which frequently occur, the changes in the tracheal wound must be mentioned.

The most frequent is diphtheria of the edges of the wound, which often become gangrenous, and give rise to large losses of substance and tracheal fistula. Unsuitable canulæ may cause superficial scabbing and ulceration of the canal of the wound, as well as of the deeper parts of the trachea. Phlegmonous and erysipelatous inflammation of the edges of the wound, and also round about it, hæmorrhage from ulceration of the anterior and posterior wall of the trachea, also occur as unfavourable complications.

As secondary local diseases, stenosis of the larynx and trachea, and hoarseness and aphonia, must here be referred to. The former are caused, as a rule, by granulations in the upper or in the lower part of the tracheal wound, or by cicatricial contraction of the healed ulceration, or by membranous adhesions and chronic thickening and infiltration of the vocal cords; paralysis, too, of the posterior crico-arytenoid muscle has been observed. The changes of the voice depend either on inflammatory processes, especially of the vocal cords, on losses of substance, or on paralysis of the muscles or nerves of the larynx.

Paralyses, beyond a doubt, occur most frequently as secondary diseases.

In general it may be held that the more severe the case, the more probable is it that these will appear. There are, however, numerous exceptions, and even slight cases are attended by them. Diphtheritic paralysis seldom appears before the second week, much oftener it appears in the third, fourth, or even the sixth. As to its nature and origin, opinions are divided. While some, and principally Ziemssen, consider that the paralyses are peripheral, others declare that they are central, and support their statement on the fact that Buhl and others have seen capillary hæmorrhages and changes in the lymphoid cells and nuclei of the nerve sheaths and ganglia of the spinal cord, and also fungus proliferations.

Pretty often there are disturbances of the power of vision from paralysis of the muscles of accommodation. These are manifested by inability to see clearly small objects near the eye; they appear dim and blurred; reading is impossible, the eyes easily become tired, the pupil is generally dilated. If paralysis attack the muscles of the eyeball, then strabismus and double vision are the consequences.

Paralysis of the soft palate and gullet are by far the most common of all. The symptoms of these are to be found in the chapter on *nervous diseases*. In severe cases paralysis of the muscles of the face occurs.

The nerves and muscles of the larynx become paralysed, sometimes together with the muscles of the pharynx, somtimes with those of the extremities.

Paralysis of the sensory superior laryngeal nerve is either unilateral or bilateral. It is one of the more dangerous complications, as patients affected by it pretty frequently die from traumatic pneumonia. Objectively it manifests itself by cessation of direct and reflex excitability.

Motor paralysis of the larynx occurs on one or both sides. It attacks individual muscles and entire groups of muscles, very often all the muscles supplied by the recurrent nerve; in the latter case the danger of traumatic pneumonia is very great. The muscles of the extremities are also often paralysed. Paralysis beginning in them makes itself known through perverted sensations in the points of the fingers and toes, a woolly feeling, formication, itching, and darting pains. Gradually weakness and defective movement set in, uncertain, staggering gait, and at last complete paralysis.

When the muscles of the throat, neck, and trunk are all affected, the patients lie in bed quite helpless, requiring to be raised up and to be fed; the head cannot be held straight, it falls backwards or forwards. When there is paralysis of the muscles of the lower part of the trunk, the patient cannot sit up in bed. Sometimes the paralysis spreads to the respiratory muscles; breathing becomes superficial, dyspnœa sets in, and the patients die asphyxiated, after that, in some cases, the muscles of the bladder and intestines have also become paralysed.

Among the somewhat rarer complications and secondary affections, which, to make the enumeration complete, must be mentioned here, are anæmia, neuralgia, sensitiveness to loud speaking, cold and heat, angina pectoris (Seitz), cramp of the extremities, hiccough, paralysis of the œsophagus, attacks of sneezing and yawning, psychical disturbances, general paralysis, idiocy, exalted state of the mind, and somnambulism (Seitz). By Pauli, acute articular rheumatism; by Cahn, disturbed assimilation; by Seitz and Wertheimber, diphtheria, with encrustation of the gastric mucous membrane, have been observed.

The diagnosis of diphtheria, in cases in which the disease is developed in some degree, presents no difficulties. That at the beginning it is possible to mistake it for other affections, must be allowed. It is most commonly mistaken for catarrh of the lacunæ and follicular angina, the differential diagnosis of which has already been discussed. It is less likely to be mistaken for pharyngomykosis benigna, aphtha and herpes, as well as for syphilis or for artificially produced eschars, especially those due to the galvano-caustic point.

The prognosis of diphtheria is always very dubious. However harmless the affection may seem at first, the practitioner should never predict a favourable issue, for too often he finds himself deceived. There is no disease less to be depended on than diphtheria, on account of its protracted course, and the unforeseen and intermediate attacks and complications which may arise.

The prognosis is comparatively favourable when the affection has attacked only one or both tonsils, the palate, or the uvula. Disease of the posterior wall makes the prognosis worse on account of the danger of the nose and larynx becoming infected. A more favourable issue may, however, be expected when the strength is good and enough nourishment is taken; of the greatest importance is the age of the patient; for adults, young people, and children above six years, a better prognosis may be given; for children under five it is less good, and for those under two years very bad. It is certain that badly-nourished scrofulous children have less power of resisting the disease, also that the prognosis to a large extent depends on the character of the epidemic. I cannot regard diphtheria of the nose as a complication which absolutely endangers life; on the contrary, it is diphtheria of the larynx, trachea, and bronchi which is the most dangerous of all complications. In the septic forms the prognosis is almost absolutely hopeless, especially in septic-gangrenous diphtheria, which only very exceptionally ends in recovery. An unfavourable issue may be expected in all cases in which the heart's action is specially impaired. The prognosis is very grave in bad cases of nephritis, or of pneumonia. Among the secondary affections, paralyses are in general, and rightly, considered favourable, although death may sometimes occur from laryngeal paralysis, or from extension of the paralysis to the respiratory muscles.

The mortality from diphtheria is highest among children under one year, being, according to Monti and Herz, about 80 per cent.; among children between one and three years it is 45 per cent.; between three and five, 40 per cent.; between five and ten, 17 per cent.; from ten upwards the rate gradually diminishes, and after twenty is very small. Against these statistics, the fatal cases mentioned by all sorts of diphtheria specialists who have cropped up, stand in the most glaring contradiction. One can now-a-days hardly ever take a publication on diphtheria into one's hand, without finding at the close the stereotyped remark, that the author has not had to lament a single fatal case since he has used this or that remedy. Without going further into the worth of such reports, which carry on the face of them the stamp of the most common advertising, I would only like to express the wish, that the gates of science may in future remain more firmly closed than before, against such sordid elements.

Treatment.—Prophylaxis is not by any means the least part in the treatment of diphtheria, and every endeavour should be made to prevent the extension of the disease. The best and only preventive remedy is isolation of the sick from the healthy, although it must be allowed that this is very often illusory, as the other children may be already infected before the order is given. Nevertheless, the healthy children should at once leave the infected house, should avoid all intercourse with it or with infected persons, and only return when the house has been sufficiently disinfected.

As to the choice of disinfectants, there is great difference of opinion. Monti recommends, instead of the almost universally-used fumes of sulphurous acid, that the floors and furniture, and everything that can be washed, be cleansed with green soap and lye, and that the hangings and pictures of the sick-room be destroyed. If this be impossible, then the walls and floors should be washed with a 1 per cent. carbolic acid solution or 1 per cent. xanthate of potassium solution; or, at least, that the room be sprayed with a solution of perchloride of mercury in water (1 to 4000).

With the same solutions, the clothes of the patient should be washed, while the patient himself, and every one who has been about the room during the illness, should have their bodies washed with carbolic soap, and take a warm bath before they again mix with healthy people. Whether the numerous prophylactic gargles which are used, containing chlorate of potash, carbolic or boracic acids, are of any use in preventing the infection, is doubtful; but, at any rate, their use can do no harm.

In order to prevent the epidemic spreading among children, it was proposed to have sanitary police regulations; as, for example, that it be the duty of physicians or relatives to prevent healthy children, out of already infected families, going to school; and that the patients be isolated in a proper hospital or institution. That the diphtheritic poison may become propagated to healthy children going to school, is certain; against Henoch's proposition, that children open to suspicion should be kept from attending school, objections have been urged so far, only on social and scholastic grounds.

When the disease has once broken out, the greatest care must be taken that the sick room is plentifully ventilated, by night as well as by day. It is also recommended that, where possible, the largest and most spacious apartment in the house be chosen as the sick room; or, still better, that two rooms be used—the one by day, the other by night. The temperature of the rooms should never be above 12° Reaumur (60° Fahr.).

A most important point is the nourishment of the patient, both as to quality and quantity. The amount of food taken is usually small, on account of loss of appetite and difficulty in swallowing; but the little that is taken should be nourishing and suitable. There are special difficulties with children, as they generally take nothing, unless compelled. Indulgence by their parents is, in such cases, most unfortunate. Of course, the nourishment must be all fluid at first, and consist of milk, strong soups, beef tea, tea, and coffee. To delay the the administration of alcohol and other stimulating remedies when the heart is weak, is certainly incorrect. A careful physician will endeavour to guard against the occurrence of cardiac failure in good time. Small quantities of wine and water, wine soups, and, if necessary, beer, may therefore be allowed from the beginning. When collapse threatens, stronger wines—Marsala, sherry, port, and champagne—may be used; in poorer practice, diluted rum or brandy, with some sugar, is to be recommended. After the membranes are shed, and the inflammatory symptoms abated, more solid food, soft eggs, tender meat, mince, fowl, and venison may be allowed.

Medicinal treatment must be both local and general. The purpose of local treatment is to overcome the inflammatory symptoms, to remove the exudation or render it innocuous, and to prevent the diphtheritic process from spreading to other mucous membranes.

For overcoming and mitigating the local symptoms, cold, in the form of cold compresses or ice bladders, applied according to the previously explained methods, is specially suitable. It must be carefully noticed that cold—outwardly and inwardly, in the form of ice-pills, etc.—must only be used until the inflammatory symptoms have abated. If the process of exudation is completed, then warm stimulating applications must take the place of cold.

The second indication, that of removing and rendering innocuous the pseudo-membranes, is best fulfilled by the careful cleansing of the throat with certain remedies, generally some disinfectant. The number of these is so great that it is impossible and unnecessary to enumerate them all. One author has got the best results with one remedy, another with another, which only goes to prove that favourable results may be had from each method of treatment, and likewise, that with each, bad results and fatal cases will be lamented.

The previously much used method of destroying the membranes by cauterising with caustic, and other similar procedures, is now in general very properly abandoned as hurtful. The more the throat is irritated and the membranes interfered with, the more certainly and quickly are they reproduced. Several authors use hot vapours for loosening the membranes; this may cause a line of demarcation to form, and so hasten the shedding. For the purpose a pot filled with boiling water or camomile tea is used, the vapour from which is inhaled, by means of a suitable funnel, as hot and for as long as possible. Steam sprays seem more to the purpose, the conducting glass tube of which is held directly in the mouth. They are preferred because by them disinfecting remedies along with heat can be brought to bear on the throat. By many a solution of carbolic acid, from 1 to 5 per cent., is valued as the most efficient of all remedies; but, according to my own experience, it is no better than any other, while the stronger solutions act as poison, and give rise to green urine and carbolic acid poisoning. For younger children a 3 per cent. solution of boracic acid is preferred; for adults and older children, a solution containing 1 to 3 per cent. of chlorate of potash. Other remedies to be mentioned are chinolin, resorcin, benzoate of soda, salicylic acid, thymol, permanganate of potash, corrosive sublimate, oil of eucalyptus, and oil of turpentine. Inhalations should be used every hour or half-hour, according to the severity of the case, and—a point of particular importance—continually during the night. They are just as useful for adults and older children, when properly and constantly carried out, as they are useless with younger and refractory patients. I have stopped using them, therefore, with the latter, and have the throat thoroughly cleansed every hour with an ordinary syringe.

The dissolving of pseudo-membranes chemically has again lately come into use. Most surprising results in liquefying fibrinous materials have, according to the coinciding reports of Rossbach, Kohts, Fräntzel, Asch, and others, been got from the application of *papayotin*, a 5 per cent. solution of which is generally used as an inhalation or as a pigment. The very high price of the drug still stands in the way of its general use.

Next to it is pepsin, which may be either inhaled or painted on.

  ℞ Pepsinæ . . . grs. 16
  Acid. mur. dil. . . . min. 4
  Aquæ destil. . . . . ʒi
Sig.—For inhalation.

As a pigment, to be used hourly, the formula is :—

  ℞ Pepsin. german. . . . grs. 16
  Acid. mur. dil. . . . min. 4
  Aquæ destil. . . . ʒiss

I prefer to use lime water with an equal part of water, and the addition of from 1 to 3 per cent. carbolic acid, and I use it sometimes for inhalation and gargling, sometimes for painting and syringing.

As remedies which dissolve membranes, lactic acid (5 per cent. solution), carbonate of soda, and carbonate of potassium have some influence. From the last two, I have several times seen good results in cases of laryngeal diphtheria.

These remedies must be used as gargles and pigments where inhalation is impossible, or along with it. That gargling is only of use in diphtheria of the isthmus has already been demonstrated.

In cauterising, it is best to use a thick, but fine-haired hair pencil, or a piece of fine sponge or wadding fastened to a wooden or whale-bone handle; the remedies to be applied must be more concentrated. The formulæ—

$$\text{R Chloral. hydr.} \quad . \quad . \quad . \quad . \quad \mathfrak{Z}\text{iss}$$
$$\text{Glycerini} \quad . \quad . \quad . \quad . \quad \mathfrak{Z}\text{i}$$

and

$$\text{R Hydrarg. perchlor.} \quad . \quad . \quad . \quad \text{grs. 24}$$
$$\text{Spir. vin. rect.} \quad . \quad . \quad . \quad . \quad \mathfrak{Z}\text{vi}$$
$$\text{Aq. destil.} \quad . \quad . \quad . \quad . \quad . \quad \mathfrak{Z}\text{i}$$

are recommended. In England, painting with a 2 per cent. solution of liq. ferri perchlor. is very much liked. Morell Mackenzie uses for this purpose, after very carefully drying the mucous membrane with blotting-paper, varnish, usually made up of alcoholic or ethereal solutions of the Balsams of Peru or Tolu (3 to 15).

The principal indication, the prevention of the spreading of the exudation, we are unable to fulfil. Every assertion to the contrary rests on false observations or is braggadocio. No one will subject his child to the method proposed by Frisch, which is that tracheotomy be performed as a prophylactic, and the larynx tamponaded to prevent diphtheria spreading to it.

The insufflation of powders is brought forward by some authors as a still milder procedure. For this, freshly powdered boracic acid is used, also iodoform alone, or with boracic acid, tannin, alum, flowers of sulphur, resorcin, chlorate of potash, benzoate of soda, or calomel,—the last after the throat is cleansed with a 5 or 10 per cent.

solution of common salt. The removal of pseudo-membranes is only allowable if they are loose in the throat, or are attached to the substratum by shreds only.

In diphtheria of the nose, cleanliness is a matter of primary importance. It is accomplished either by means of a spoon, the nasal bath, or a specially constructed syringe. I myself use for children an india-rubber bag, with a point made either of bone, well-rounded, or of india-rubber. The fluid to be employed must be lukewarm, and, with the nose completely closed up, should be injected every two or three hours by a strong squeeze of the bag. Particularly suitable are —lime-water with carbolic acid or spirit, and boracic acid (2 per cent.), chlorate of potash (2-3 per cent.), lactic acid, carbonate of soda, and in case of fœtor, permanganate of potash (½ per cent.) solutions. When hæmorrhages occur, the injection of solutions of tannic and carbolic acids, alum, or liq. ferr. perchlorid. are preferred. After cleansing, powdered boracic acid, or iodoform and carbonate of magnesia (equal parts), are insufflated. To protect the skin from the excoriating influence of the secretions, the application of boracic acid in vaseline is recommended.

The treatment of laryngeal diphtheria by means of cold has not proved beneficial, since it neither retards the development of the disease, nor in any way influences the already-developed affection; also from energetic inunction with mercurial ointment, recommended by Bartels, nothing is to be expected. Although the benefit derived from warm stimulating applications to the neck, and inhalations of hot medicinal vapours, and especially from the already-mentioned solvents of false membrane, is usually nothing extraordinary, still they should always be made use of in lieu of a better therapy. Emetics, which in laryngeal croup are so much liked, are in diphtheria not only utterly useless, but actually injurious.

If breathlessness increases, then removal of the membranes is indicated. With older children and adults it may be attempted, under the guidance of the laryngoscopic mirror, to remove mechanically the membranes lining the larynx and trachea, by means of a fine hair pencil, or small piece of sponge fastened to a German-silver wire, impregnated with lime water, carbolic acid, papayotin, or pepsin. A rotatory movement must be imparted to the instrument. That these attempts are not altogether hopeless, I have myself often experienced, being once successful in removing a well-formed cast of the larynx and trachea, and thus avoiding the necessity for performing tracheotomy. Also by means of Schrötter's dilatation tube, of hard india-rubber, which I use for tubage of the larynx, and which, if necessary, I introduce into the larynx, guided by my finger, the breathing can for a little while at least be facilitated. Since all these procedures cannot be carried out with children, and even with adults are very often without result, only one other remedy still remains, viz., tracheotomy.

The operation is indicated when there is continual dyspnœa, and when the thorax and diaphragm begin to be drawn in; to wait any longer is death to the patient. Without going into the technicalities of tracheotomy, it may here be specially mentioned that, although the results of the operation are not so brilliant as in real laryngeal croup, still a third, or even the half of those operated on may be saved, and that it is the duty of every conscientious physician, to draw the attention of the relatives to the operation as the only means of saving the patient's life, and also to perform the operation.

The after-treatment is quite as important as the operation itself. The wound of course must be covered with carbolic or iodoform gauze, or with boracic lint, and treated with strict antiseptic precautions, the atmosphere of the room must be kept constantly moist by means of open vessels containing water, the patient must be made to inhale through the canula every hour solutions containing 1 per cent. of sodium bicarbonate and carbolic acid, and the canula must be carefully cleaned, and no pieces of dry secretion or pseudo-membrane allowed to lodge in it.

The general treatment of diphtheria is quite as important as the local.

Physicians have for a long time been striving to find a specific remedy for diphtheria, *i.e.* one that will, on the one hand, remove the exudation and destroy the virus, and, on the other hand, prevent the extension and the new formation of pseudo-membranes. Such a remedy has not as yet been found, but there is no want of substances which, it is believed, exert an influence on the diphtheritic process. Of these chlorate of potash is specially valuable, its dose being grs. 8 to ʒi of water for children, and grs. 16 to ʒi for adults, a tea or tablespoonful respectively every hour; but great caution is necessary, as poisonous symptoms may easily occur. Great hopes are placed on the muriate of pilocarpine, from which Guttmann has seen good results. He prescribes it as follows :—

  ℞ Pilocarp. muriat. . . . grs. ⅛–¼
   Pepsinæ . . . . grs. ⅓–½
   Acid. mur. dil. . . . . min 2
   Aq. dest. . . . . . . ʒi
  Sig.—A teaspoonful every two hours.

Along with other observers, I consider it useless, and when the heart is weak, as very dangerous.

Corrosive sublimate also is looked upon as a specific. It is given thus :—

  ℞ Hydrarg. bichlor. . . . . gr. ⅕
   Albi ovi . . . . . . No. 1
   Aq. dest. . . . . . . ʒi
  Sig.—A teaspoonful every hour.

I personally put aside all specifics, and combat the individual symptoms. For fever in children I give salicylate of soda (15 to 45 grs. daily) in solution. In the septic forms I use the muriate of quinine, or the citrate of iron and quinine (grs. 8 to 15 daily), recommended by Monti in preference. As a tonic, from the first I order for children a teaspoonful every two hours of—

  ℞ Liq. ferri perchlor. . . . . ʒiss
   Aq. dest. . . . . . . ʒi
   Syr. simpl. . . . . . ʒiss
  Sig.—A teaspoonful every two hours.

which is also used in England, or the Tinct. ferri acet. æth. Collapse occurring suddenly, must be met with warm baths, the warm pack, alcohol, liquor. ammon. anis., or the anodyne liquor of Hofmann, and, if necessary, camphor injections. Threatened uræmia is treated with digitalis or stimulants by Leube, according to the condition of the pulse. Nephritis calls for diuresis, to be brought on by means of effervescing drinks, milk diet, and alum administered internally. Pneumonic infiltration or bronchitis require expectorants; for diarrhœa opium must be given. In myocarditis every movement of the body must be strictly forbidden, and, as in anæmia, iron, alcohol, and nourishing food, as well as country air, are recommended.

Against diphtheria of the trachea and bronchi, every remedy is generally powerless. Attempts must always be made to loosen the membranes by aspiration, by means of a catheter or sucking syringe, and by instillation and inhalation of papayotin, pepsin, or lime water, and to hasten their shedding by expectorants (Inf. senegæ with Liq. ammon. anis.).

The affections of the sexual organs and those of the skin, also diphtheria of the tracheal wound, must be treated on surgical principles, by the disinfecting remedies already discussed.

Although, as a rule, diphtheritic paralyses get well of themselves, it seems necessary and desirable to hasten the cure by therapeutic means, particularly in the more dangerous forms of laryngeal paralysis, paralysis of the œsophagus and of the extremities. The sovereign remedy is electricity, the constant as well as the induced current. I have obtained wonderfully quick results in pharyngo-laryngeal paralysis by endopharyngeal faradisation and galvanism alternately applied. For the more obstinate forms Ziemssen recommends subcutaneous injection of strychnine as a reserve remedy.

Granulations in the trachea must be removed, and membranous adhesions of the larynx or tracheal wound separated by means of a knife or the galvanic cautery, while cicatricial stenosis is to be removed by Schrötter's tubes, and laryngeal catarrh treated locally.

GANGRENOUS PHARYNGITIS.

Synonym.—Pharyngitis gangrænosa.

As in the mouth, so also in the pharynx, gangrene comes under observation as a modification or result of the different forms of inflammation.

Although it occurs most frequently in consequence of diphtheria, yet it may appear independently as a primary disease. It is either circumscribed or diffuse, and generally attacks weakly, badly-nourished cachectic children and adults. It occurs along with phlegmonous angina and retropharyngeal abscess, or with the secondary anginæ of scurvy, scarlatina, measles, typhoid fever, and variola, or develops after injuries and operations, or after hæmorrhages into the mucous membrane. Pitha observed an epidemic appearance of angina nosocomialis phagedenica in hospital gangrene, as also did Browne. According to Barthez and Rilliet, pharyngeal gangrene may also occur in consequence of whooping-cough and pharyngeal tuberculosis. Semon has observed localised gangrene on the tonsils.

The disease usually begins in primary cases on the soft palate and tonsils, more rarely on the posterior wall, with symptoms of a catarrhal or phlegmonous inflammation, with fever and severe local symptoms. After some time, generally two or three days, the first symptoms of necrosis appear in the form of brownish black spots, over which the mucous membrane is elevated in the form of bullae. Their contents at first are serous, later on discoloured, sanguinolent, and ichorous. At the same time the lymphatic glands swell up, and the affection runs its course from this time onwards usually in the form of adynamia. The fever falls, the temperature becoming normal or subnormal, the pulse slows, sometimes to as low as fifteen beats per minute, the wave is small, the skin pale, cold, cyanotic, and covered with sweat. In place of the bullae there appear dirty, blackish grey masses, which bleed easily, and which, on separating, leave behind them deep losses of substance and perforations, and often complete destruction of the soft palate.

In consequence of the local symptoms, of want of appetite, of the often very profuse hæmorrhages, or of pyæmia, patients generally die in about eight days. Consciousness may remain, but is often very dull, and coma and stupor set in. The odour from the mouth is very penetrating, the secretion copious, bloody, and ichorous, and sometimes petechiæ of the skin and hæmorrhages of internal organs occur. Gangrene has also repeatedly been observed in other parts of the body—the lungs, the larynx, the generative organs, especially of women, and also in the parts surrounding the anus.

The prognosis of pharyngeal gangrene, whether primary or secondary, is very bad, the disease being almost always fatal, though cases of cure have been recorded by Musset, Trousseau, and Mackenzie.

Treatment.—Treatment is very powerless, still the attempt to localize the gangrene, and to mitigate the local and general symptoms, must not be omitted. The remedies used are the same as in oral gangrene and pharyngeal diphtheria.

SYPHILIS.

Among the secondary affections of the pharynx, none is of greater importance than syphilis.

Apart from the horrible cases of primary infection, the pharynx only becomes affected, when the syphilitic poison has found its way into the lymphatic and blood vessels.

The first manifestation of the disease is erythema, known as syphilitic angina. Generally, along with roseola of the skin, there appears on the mucous membrane of the soft palate, tonsils, and posterior wall, a patchy or diffuse uniform, more or less pronounced, redness and slight swelling, which runs a chronic course, and is accompanied by very trifling symptoms. The differential characteristics of syphilitic angina—viz., dusky red colouring of the mucous membrane, uniform redness sharply limited to the soft palate, are not satisfactorily seen; a more certain diagnosis can only be made when other syphilitic symptoms are present. Cloudiness of the epithelium is always suspicious, as it usually occurs in syphilitic inflammation of the mucous membrane, but is not demonstrative.

An almost daily occurring symptom, is the appearance of mucous patches or papules, and ulcers of the pharynx arising from them. These are generally observed along with the same appearances in the mouth, but the throat may be attacked first and alone.

Their favourite situation is the soft palate; they also occur on the pillars of the fauces, the uvula, and the tonsils, and comparatively rarely on the posterior wall.

On normal, but generally more or less hyperæmic and swollen, even œdematous mucous membrane, examination reveals milky bluish white spots and streaks, varying in size, round or oval in shape, slightly raised above the mucous membrane, and covered by a loose macerated epithelial covering. If this covering becomes thrown off, then there are seen superficial longitudinal ulcers, from one to several centimetres in size, often irregularly formed, which make the free edges of the pillars of the fauces appear serrated and grooved. These notches are seen very clearly at the base of the uvula and on the anterior pillar. On the tonsils the patches are either isolated or uniformly distributed over the hypertrophied tonsils, or they form confluent spots and superficial ulcers.

The symptoms are generally slight, or are identical with those of acute angina. Pain on swallowing, especially when sour and highly seasoned foods are taken, as well as from smoking and speaking, is seldom absent.

In the later stages of syphilis, syphilitic nodes, also called gummata or syphilomata, and the ulcers they give rise to, are, on account of their unusual frequency, of special interest.

The syphilitic node appears either diffusely as an infiltration, or circumscribed as a gumma. In the naso-pharynx they occur most frequently on the salpingo-pharyngeal fold, and are seen as broad, smooth, roundish, yellowish-red, or red swellings, from the size of a pea to that of a hazel-nut. Their course is less characteristic on the posterior wall of the oral and laryngeal parts of the pharynx. Sometimes towards the middle, sometimes altogether at the side, behind the posterior pillar, painless or almost symptomless circumscribed red swellings develop, which often bear a striking resemblance to heaped up granulations, or hypertrophied lateral bands. Their true character seldom becomes unfolded, till they decay and change into ulcerations.

Gumma of the soft palate appears in quite an unmistakeable manner.

At its commencement it generally escapes observation, because the syphiloma occurs as a rule on the posterior surface of the palate : it enlarges, however, and reaches the anterior surface, or may even develop there first, and then appears as a circumscribed bluish-red deposit, arching forward the mucous membrane : the parts round about are markedly hyperæmic, and the whole of the palate and uvula are œdematous. As it goes on, white or yellowish streaks and specks appear on the nodes, a sure sign of commencing softening, which, gradually increasing, often lead to perforation, sometimes with unexpected rapidity. At first the opening corresponds more or less to the circumference of the softening deposit; later on, however, it increases, especially if nothing is done for the patient. Sometimes there are several nodes on the soft palate, which simultaneously or one after the other break down, and coalescing, give rise to extensive ulcers. Complete loss of the soft palate is not uncommon, but generally only a part is destroyed. When the nodes develop near the base of the uvula, it may become entirely separated, or may be kept attached by a thin band. Adhesion of the soft palate to the posterior wall may occur, or ring-shaped stenosis, and membranous adhesion of the lower section of the pharynx; these, however, have been already mentioned. After the deep ulcers heal, white shining cicatrices are left, which cause contraction of the neighbouring parts and give rise to the previously mentioned changes. The formation of cicatrices on the palate interferes with its movement and contractility, or abolishes them altogether, and thus renders the closure of the naso-pharynx impossible, from the abnormal shortness of the palate.

When gummata occur on the tonsils, these organs swell up till they touch, and still further till they become flattened against each other. The superficial situation of the nodes makes the tonsils appear rough and tuberculated; after a while the nodes soften and leave deep cup-shaped ulcers, which run together and convert the tonsils into a ragged, corroding, often granulating mass. There often develop in the depressions and lacunæ, white streaks crossing each other, which consist partly of cicatricial tissue and partly of new proliferating connective tissue (interstitial retracting tonsillitis).

The subjective symptoms of syphilitic ulceration of the pharynx manifest themselves mainly during deglutition. The act of swallowing is not only painful, but also so far prevented, in that hard pieces of food cannot be, or are only with difficulty, swallowed. The timbre of the voice is usually nasal and muffled. When the infiltration of the palate is very great, and when the palate is perforated or considerably destroyed, regurgitation of food takes place.

The diagnosis of mature pharyngeal syphilis, but especially of ulcers, is generally easy, and depends on the changes described, and other co-existent symptoms or traces of syphilis. That it should be mistaken for diphtheria or catarrhal ulcers appears scarcely possible, but proves that the recognition of a disease so widely spread as syphilis, has not yet become general. I can only advise young practitioners, on examining an ulcer in the pharynx, always to look upon it as the manifestation of a dyscrasia, in the first place as syphilis, then as tuberculosis, lupus, or carcinoma.

Sometimes it is very difficult to differentiate between syphilitic patches and follicular angina, since both may be accompanied by fever. But the diagnosis becomes much more difficult when, as happened in one of my cases, an exanthema is also present; then at least confounding it with measles or scarlatina can hardly be avoided, but the continuance of the rash after the fever is gone, and the fever never being so great in syphilis as in an acute exanthema, show with certainty that it is syphilis. Herpetic and aphthous ulcers are also often diagnosed as syphilitic. The acute, generally febrile course, the sudden development of the ulcers, the severe pain and rapid healing, are against their being syphilitic; while, on the other hand, the gradual commencement, long duration, change in form and situation, the absence of fever, and slight pain, point to their being syphilitic.

The prognosis is favourable, provided the patients do not come under treatment too late, or with irreparable losses of tissue.

Treatment.—The treatment must be both general and local. It would be unscrupulous to prefer the one method at the cost of the other; the best and quickest results are obtained by combining the two.

Mercury is the most suitable remedy for early syphilitic symptoms, whether administered by inunction, by injection, or internally. For gummatous forms I know no prompter and better remedy than iodide of potassium. I have often succeeded in making already completely softened gummata disappear, and in preventing perforations, by its energetic administration.

Local treatment consists of gargling, inhaling, painting, and cauterising. The beneficial influence of chlorate of potash is well seen here. Patches do well on being painted with nitrate of silver solutions, but are more quickly influenced by using it solid. For cauterising gummatous ulcers I can best recommend Mandl's solution with carbolic acid. Deep ulcers, and the edges of perforating ulcers of the palate, must be thoroughly cauterised with solid nitrate of silver; it often succeeds in reducing the perforations to a minimum, or at least so far as to prevent the voice becoming affected.

Larger, especially osseous, perforations, and complete destruction of the soft palate, can be covered by palate plates and similar apparatus, or may be remedied by surgical operations; restoration of normal speech, however, is very seldom obtained.

TUBERCULOSIS.

The localisation of general tuberculosis in the pharynx is comparatively rare. Considering its extraordinary frequency in the larynx immediately adjoining, it must appear very remarkable how seldom the pharynx participates; unfortunately science has not yet answered the question: What is the origin of this disease?

The question, whether tuberculosis of the pharynx is, or is not, a primary disease, is raised in this case as well as in that of the larynx; the possibility that the tubercular virus is first localised in the pharynx must be granted; that the pharynx may be attacked earlier than the lungs and larynx, I myself have often seen. My observations always force me to the conclusion, that the pharynx is only apparently primarily attacked, by which I mean, that already, before the outbreak of pharyngeal tuberculosis, cheesy or tubercular deposits exist almost without exception in other organs, although objectively they cannot always be proved to exist. The virus may, as in syphilis, lie dormant for years, and then suddenly develop its deleterious effects.

As proof of this, I may here mention that one of my patients, when twenty-eight years old, suffered from an apex catarrh with hæmoptysis; after fifteen years of perfect health he was suddenly seized with purulent inflammation of the middle ear and extensive tuberculosis of the soft palate; while another, who, when sixty-three, had his right testicle excised for tubercular degeneration, was attacked two years later by tuberculosis of the posterior wall of the pharynx, and died from the disease extending to the larynx and afterwards to the lungs.

Tuberculosis of the pharynx generally accompanies pulmonary tuberculosis, the presence of which has been already demonstrated. The disease most usually occurs between the ages of twenty and fifty, and mostly in the male sex; **very** seldom in children. Only Isambert, Schepelern, Millard, and Stoerk, mention such cases.

The changes which take place, are all due to tubercular infiltration of the mucous membrane. The tubercles, as in the larynx, are generally situated under the epithelium, and therefore invisible to the naked eye; often, however, they become distinctly recognisable to the unaided eye, on the mucous membrane itself, and are seen as small grey or greyish-yellow nodes of the size of millet seeds; their favourite situations are the soft palate, posterior wall of the pharynx, tonsils, and naso-pharynx, where, according to Eugen Fränkel, Wendt, and Zawerthal, the pharyngeal tonsil, the parts surrounding Rosenmüller's fossa, and the fornix, are specially liable to be attacked. According to Strassmann, tuberculosis of the tonsils is a very common accompaniment of pulmonary tuberculosis, but does not, however, produce symptoms.

Pharyngeal tuberculosis seldom comes under observation while the nodes are forming, but generally only when ulceration has produced softening and desquamation. According to the excellent description of B. Fränkel, who was the first in Germany to call attention to this disease, tubercular ulcers of the pharynx are lenticular in shape; they are more superficial than deep, and have a pale yellow base with irregular eaten-out edges, and are very often covered with small pale red granulations, and with thin unhealthy pus; round about, or on the edges of the ulcer there are often papillomatous excrescences and warty growths; the lymphatic glands of the neck are usually swollen.

Among the subjective symptoms, pain takes the first place. It is extremely violent, particularly during deglutition, and radiates to one or other ear by transmission of the painful sensations through Jacobson's nerve, the glosso-pharyngeal, and the auricular branch of the vagus. Pain on swallowing is generally so intense, that patients rather suffer hunger and thirst, all the more if the food when taken is thrown up again through the nose. Secretion of mucus is increased, there is constant irritation, causing hawking and swallowing, and the breath has a very bad odour. Besides these local symptoms, there are also loss of appetite, emaciation, debility, evening fever, usually of a very irregular type, quickening of the pulse, and perspirations, and cough and hoarseness, especially if the lungs and larynx are also diseased.

After a longer or shorter time the primary tuberculosis of the pharynx spreads to the larynx, the mouth, and the lungs, or to the brain and intestines, and in from three to ten months causes the death of the patient.

The diagnosis sometimes presents great difficulty, not only to the young, but also to the experienced practitioner. It is most liable to be mistaken for syphilis. Such an error may be avoided, if one considers, that tubercular ulcers of the pharynx, in contra-distinction to those of the larynx, never penetrate so deeply as syphilitic ulcers do, that they possess a distinctly atonic character, and that as a rule other manifestations are also present, such as tubercle of the testicle, rectal fistula, or pulmonary phthisis. If the latter are absent, the diagnosis must remain in suspense, or be made *ex juvantibus*. Microscopic examination revealing tubercle bacilli makes the diagnosis certain. Further, I can agree with B. Fränkel when he says, that whoever sees pharyngeal tuberculosis even once, will never again forget its appearance.

The prognosis is absolutely hopeless, and if some—among them Gougenheim and Küssner—speak of healing the disease by means of iodoform and cauterisation, it seems more than likely that doubt may be thrown on the correctness of the diagnosis.

Treatment.—Therapeutics are entirely powerless, and must be confined to cleansing the ulcers, relieving the pain, and keeping up the strength. To prevent the unbearable pain on swallowing, insufflation of morphia is most recommended, combined with starch or boracic acid, the proportion being morphia 1, starch powder 10 parts. For disinfecting and removing the bad smell, insufflation of iodoform is used, but I prefer painting with creasote (1) in glycerine (100). Besides these, cleansing gargles and inhalations of 4 per cent. boracic acid, ½ to 1 per cent. carbolic acid, or chlorate of potash, grs. 15 to ʒi, are of some benefit. It is easily understood that in so painful a disease, the greatest care must be observed in respect to the consistence, temperature, form, and unirritating nature of the food.

SCROFULA, LUPUS, LEPRA, AND GLANDERS.

In the course of one's reading, one now and again meets with cases of scrofula of the pharynx.

Thus, amongst others, Homolle speaks of a tubercular and of a purely ulcerating form of pharyngeal scrofula. By the first he means lupus of the pharynx; in the ulcerating form of scrofula, adhesions in the pharynx do not commonly occur; on the other hand, perforations of the soft and hard palate may be exclusively the result of strumous ulceration. Considering these observations from an impartial point of view, it appears to me, however, much more likely that the diagnosis should be syphilis. The differential diagnostic characteristics, which Homolle brings forward,—that syphilitic ulcers have generally sharp edges and a granulating base, and that scrofulous have gradually sloping edges and an atonic base, that the parts round about syphilitic ulcers are reddened and swollen, and that those around scrofulous ones are pale,—are too inconstant, and too negative to found a diagnosis upon.

Wendt mentions that there is found on the posterior wall of the naso-pharynx, and also on the pharyngeal tonsil, in scrofulous and tubercular persons, as well as in those who are to all appearance healthy, a tubercular cheesy scrofulous inflammation of the follicles—*pharyngitis scrofulosa*. The follicles are generally very numerous, of a greyish-yellow, or yellowish colour, and of a dry cheesy consistence, or are surmounted by a mass of fatty detritus. By the decay of the outer surface small ulcers are formed, and when the more deeply embedded ones necrose, numerous round defects are left in the parenchyma; on several neighbouring ulcers coalescing, considerable loss of substance and adhesions of the soft palate may occur. B. Fränkel, Lewin, and Michel also record the occurrence of scrofulous ulcers of the pharynx.

Until more exact observations are made, it is as well to be suspicious of scrofulous ulcers of the pharynx, since, as E. Wagner has rightly remarked, they have almost always a tubercular, syphilitic, leprous, or lupoid origin.

It is certain that *lupus* occurs most frequently in scrofulous persons. It is met with both primarily and secondarily; secondarily along with lupus of the outer skin, but more often, however, with that of the nose and of the larynx. It attacks principally young females, its favourite seat being the tonsils and soft palate. The tonsils appear enlarged, their outer surface rough, and covered with excrescences, between which ulcerations are seen. If the palate is attacked, then, according to Krause, smaller or larger confluent nodes develop on the mucous membrane, which is either normal or is pale. These nodes either shrivel up, or break down and form deep ulcers, which either cicatrise, or give rise to defects and perforations or to adhesions and other anomalies.

The diagnosis of lupus is sometimes difficult, as it strongly resembles syphilis. The papillary-nodular appearance, the very slow and painless course, the presence of lupus of the skin, and other scrofulous appearances, as well as the youth of the patient, all point to lupus.

Treatment.—The treatment of lupus must be both general and local. As in some cases it rests on a syphilitic basis, a trial of iodide of potassium is indicated; lupus in scrofulous persons is, however, not generally influenced by it. For them are recommended suitable diet, country and sea air, quinine, iron, and cod liver oil, to improve nutrition and the state of the blood. Locally, all diseased parts are to be destroyed by cauterising with a solution of liq. fer. perchlor. (1 to 4), or by applying nitrate of silver, or the galvanic cautery, and scraping with curettes and sharp spoons.

Lepra of the pharynx is one of the rarest diseases in Europe.

It attacks the throat only in the tubercular form, and, according to Ramon de la Sota, never occurs primarily, but only when the skin is also affected. The eruption in the pharynx is preceded by a bright redness; the leprous tubercles are white, soft, of various sizes, sometimes scattered over the swollen mucous membrane, sometimes grouped together in circles, and either normally sensitive or quite insensitive. The nodes in lupus are clear red or dark red, hard, and larger than those of lepra. The ulcers which result from the decay of leprous tubercles are soft, eaten-out, and insensitive, while those of lupus have hard turned up edges and an excavated base, are slightly painful, and leave behind scars with normal sensibility.

The prognosis of lepra is absolutely hopeless.

Treatment.—The treatment consists in change of air, nourishing diet, cold water cures; inwardly creasote (min. $^2/_5$ in pill) and salicylate of soda (grs. 30–75 per diem) may be tried. Local treatment is the same as in lupus.

Glanders, and *foot and mouth disease*, also sometimes occur in the pharynx. The forms in which they appear, are the same as those described under diseases of the mouth.

PARASITIC DISEASES.

The pharynx as well as the mouth may be affected by lower organisms.

The pharynx is seldom affected alone, more frequently in common with the mouth, hence the parasites attacking it are the same as those of the mouth.

It was previously pointed out, that in children, *thrush* very often spreads to the soft palate and posterior wall of the pharynx; in adults, pharyngeal thrush is found only in serious and wasting diseases, occurring most frequently in the course of very severe types of disease. Damaschino, among others, described a thrush epidemic among the patients suffering from typhoid in Paris. The mycelium of the thrush fungus penetrates not only into the epithelium, but also into the muscular layers. According to Duguet, who also made his observations in Paris, thrush in typhoid patients spreads from the pharynx to the mouth, while in children the opposite takes place.

The complaints of the patients are limited (if they are not drowsy, or show no other brain symptoms) to burning, lancinating pains in the neck, and inclination to vomit. In children, difficulty in swallowing and regurgitation of food occur, when the fungus is developed to any great extent.

The diagnosis in most cases can only be made by means of the microscope.

The treatment is the same as in thrush of the mouth. French observers recommend borax and Vichy water.

Another affection due to a fungus, is that which was first observed by Bernard Fränkel in 1873, called *Mycosis tonsillaris benigna*, which is identical with the *Pharyngo-mycosis leptothricia* of Hering.

By these terms is understood a disease of the tonsils and base of the tongue, which manifests itself in whitish or yellowish-gray soft, sometimes horny and often pedunculated tubercles or thorn-like excrescences. They are situated on the surface and in the lacunæ of the tonsils, more rarely on the pillars of the fauces and posterior wall, and the deep parts of the lateral folds are sometimes occupied by them. On the tongue they are situated without exception near the circumvallate papillæ, and form there filaments and fasciculi two to eight millimetres long, and up to two millimetres thick. The tonsils are either affected alone or along with the base of the tongue. Women appear to be more often affected than men, and usually between twenty-eight and thirty-five years of age.

The subjective symptoms generally consist in itching, dryness, burning, slight pain on swallowing, and tendency to hawk and cough; in some cases—as in one of my own—fever with general debility, lassitude, and loss of appetite, precedes the outbreak of the spots; in others there are no symptoms whatever, and the patients only become aware of their condition on examining their throats. If the case is very protracted, the appetite may suffer, and dyspepsia and emaciation occur, either in consequence of unsuitable interference, or on account of a hypochondriacal state of the mind.

Microscopic examination of the spots just mentioned, reveals the presence of very numerous micro-organisms, the naming of which is still a matter of dispute. In the case communicated by B. Fränkel, and in the one observed by myself, the tubercles consisted entirely of bacilli and roundish cocci; in that published by Eugen Fränkel and analysed by Sadebeck, a peculiar bacillus was found, which, on account of its appearing in tufts, was called *Bacillus fasciculatus*. Hering, who examined six cases himself, came to the conclusion, already come to by Klebs, that the discovered fungus mass consisted of leptothrix, especially as the filaments and bacilli were coloured blue, when a solution of iodine was applied to them.

The course of the affection is very obstinate, as whenever the masses are removed by operation, they immediately become reproduced.

To one who sees the affection for the first time, the diagnosis is not easy; he nearly always mistakes it for diphtheria, especially when general and well marked local symptoms are present; but against diphtheria are the usually non-febrile course, the absence of inflammatory changes in the mucous membrane, and of swelling of the lymphatic glands, the scattered appearance and consistence of the deposits, and also the simultaneous affection of the tongue. Mistaking it for angina follicularis is prevented by the always febrile and acute course of the latter, as well as by the crumbling consistence of the spots. Concretions in the tonsils are differentiated from parasitic deposits by their stony, chalky, hardness, by their being composed of fatty epidermal flakes and the component parts of chalk and cholesterine crystals, and by their exhibiting a much smaller number of bacteria.

Treatment.—Treatment must be very energetic. All observers, however, agree that the destruction and removal of the fungus deposits are impossible with the ordinary antiseptic remedies. Their removal, therefore, by mechanical means with sharp spoon, forceps, or the galvanic cautery, is recommended. If the disease is confined to the tonsils, then tonsillotomy is the simplest and most radical method.

In persons who do not feel disturbed by the presence of these perfectly harmless parasites, no treatment need be applied, as the affection generally goes away of itself, particularly if smoking be greatly indulged in.

Sarcinae may also appear to an excessive degree in the pharynx, and thence spread to the lungs; such a case having been recorded by C. Nauverk.

HÆMORRHAGES.

Hæmorrhages on the surface, and into the tissue of the mucous membrane of the pharynx, occur but rarely.

Copious hæmorrhages occur most frequently after injuries, especially in children when the soft palate has been wounded by some object or foreign body held in the mouth, and also after surgical operations, particularly tonsillotomy. Slight capillary hæmorrhages very often follow violent coughing and hawking, especially when there is chronic congestion of the mucous membrane, or when granulations, adenoid vegetations, or ulcerations and gangrene are present. Hæmoptysis, apparently originating in the throat, usually occurs in consequence of hæmorrhage from the posterior nares.

Ecchymoses, extravasations, and hæmatomata frequently accompany acute anginas, and also diphtheria; some of the less frequent causes are scurvy, purpura rheumatica, and morbus maculosus of Werlhoff. In effusion of blood into the uvula (Staphylhämatom), which arises from contusions, unskilful handling of instruments, or from swallowing large hard substances, the blood gravitates downwards, and imparts to the uvula a club-shaped appearance.

The prognosis depends on the cause and degree of the bleeding. Hæmorrhage from tonsillotomy has already been discussed. Narrowing of the pharynx, and difficulty of swallowing and breathing, may be caused by submucous hæmorrhages, which may also lead to gangrene.

Treatment.—The treatment of unimportant hæmorrhages consists in gargling, or inhaling, or spraying with ice water, or with 2 to 3 per cent. solutions of alum, tannin, or liq. fer. perchlor. Cauterisation with solid nitrate of silver is particularly efficient in bleeding ulcerations; in more profuse hæmorrhages digital compression is indicated, or pressure with cotton wool tampons impregnated with some styptic fluid, which must be kept fixed on the bleeding spot as long as possible, undisturbed by the movements of swallowing or vomiting. As already mentioned, the hæmorrhage caused by injury of the carotid can only be stopped by ligature of the artery. Hæmatomata do not as a rule require any special treatment; if from their size they cause inconvenience, their contents must be evacuated.

Hæmorrhage into the retro-pharyngeal connective tissue—*hæmatoma retro-pharyngeale*—has been observed by Stoerk.

The origin of this lesion was traced to pressure on a goitre of long standing; hyperæmia, from the neck tie being too tight, is often occasioned, not only in the thyroid gland itself, but also in the vessels of the pharyngeal wall anastomosing with the thyroid artery, which may lead to extravasation into the submucous tissue. The lesion usually appears quite suddenly, sometimes in a few minutes, or it may develop gradually.

Treatment.—Treatment can only be operative. Stoerk opens the swelling by means of a bistoury; from it there sometimes squirts out a stream of bright red blood, making one almost think one had opened into an aneurism; after opening, antiseptic gargles and inhalations should be used.

FOREIGN BODIES AND CONCRETIONS.

Foreign bodies in the throat are of great practical interest, on account of their frequent occurrence. The objects which most often stick in the throat are fish bones, tooth brush bristles, straws, ears of corn, coins, needles, false teeth, buttons, fruit stones, &c.

Very rarely do they stick in the naso-pharynx. Cases, such as that recorded by Urbantschitsch, in which the branch of a panicle of oats made its way along the Eustachian tube into the middle ear, and thence to the external auditory meatus, are very unusual indeed.

The most common occurrence is for some pointed object to stick in the isthmus of the fauces, or on the tonsils. As yet I have only found fish bones here; they often penetrate so deeply that only a very small piece can be seen; if they are very large, they may lie right across the pharynx. Besides the tonsils, the lateral walls of the pharynx, and the pyriform sinuses, are the places where foreign bodies are generally retained. Larger roundish objects generally remain behind the arytenoid cartilages, or stick between the tongue and the epiglottis.

The symptoms to which foreign bodies in the throat give rise are caused by the nature, size, and situation of the objects. Sharp pointed objects excite a continuous feeling of violent darting pain on swallowing and speaking; they cause injury to the mucous membrane, with secondary inflammation, hæmorrhage, œdema, ulceration, and abscess; and by opening into an artery may cause a fatal result from hæmorrhage. Larger objects, which remain sticking in the laryngeal part of the pharynx, besides causing difficulty in swallowing, give rise to breathlessness, and even to suffocation, from pressure of the epiglottis on the aditus laryngis, or from œdema.

Foreign bodies may be retained in the pharynx and larynx for an inconceivably long time, without exciting any special symptoms. As a rule, however, anxiety drives the patient to the physician, with the request that the object be removed as soon as possible, which, of course, is the most rational thing to be done, after making sure the object is there.

As very many errors are made in the diagnosis and treatment of foreign bodies, I shall here mention shortly the most important points.

Treatment.—When a patient comes to the doctor with a foreign body in his throat, he must briefly inform him of its nature, size, and probable situation. Although the sense of locality in the throat is not great, still, as a rule, the side on which the object is sticking can be correctly indicated. A careful inspection of the throat, and a laryngoscopic and rhinoscopic examination must now be made. If by these means one does not succeed in seeing the object, it is probable, as happens in nearly half the cases, that the foreign body is no longer in the throat, having been swallowed or coughed up; in some cases the patient for a long time feels as if it were still there, in consequence of the injury it has caused. But before this conclusion is come to, do not forget that two foreign bodies may have entered the throat, of which only one has been removed, while the other is still present; and do not in any case neglect to palpate every part of the pharynx thoroughly with the finger, and to probe the sinus pyriformis. I have very often succeeded in discovering and removing foreign bodies, which have been deeply embedded in the mucous membrane of the sinus pyriformis, touch being more certain than sight.

The removal of sharp pointed objects is best accomplished by means of a bent laryngeal forceps. When they are in the naso-pharynx, a long pair of ordinary forceps, or dressing forceps, does very well. Roundish objects, such as glass beads, coins, and buttons, when in the sinus pyriformis, must first be loosened with a bent hook-shaped probe, and brought into another situation before they can be extracted with forceps. When soft objects, such as pieces of food, bread, vegetables, or meat, stick in the upper part of the œsophagus, it is best to push them down into the stomach with the pharyngeal catheter; it is unjustifiable, however, to push hard objects with corners directly into the œsophagus; here, under all circumstances, the flexible forceps, or the coin catcher, must first be brought into requisition. If symptoms of suffocation are threatening, one must not put off time with attempts at extraction, but proceed at once to perform tracheotomy. When the danger to life is removed, the search and removal of the foreign body may be proceeded with calmly and at leisure. If no threatening symptoms appear, extraction may be postponed, particularly as many foreign bodies often become loosened without artificial help, and are ejected by coughing.

Concretions which are situated in the tonsils, give rise to symptoms similar to those excited by foreign bodies, viz., pressure, stinging pain, and difficulties in swallowing. After repeated attacks of follicular angina, there remain, as already mentioned, saucer and bowl-shaped holes and depressions, which are either on the surface of the tonsil, or correspond to the entrances of the lacunæ. These depressions are very often divided into several compartments by cord-like cicatrices and bridges, and in them pieces of food lodge, and there calcify. Best known are the tonsillar plugs or thrombi, those yellowish-green, yellowish-white lumps, having a most offensive odour when rubbed, which, by anxious persons, are presented to the physician for examination as coughed-up pulmonary tubercle or diphtheritic membranes. By the precipitation of lime, they are changed into *tonsil stones*, which reach from the size of a pea to that of a walnut. They consist of carbonate and phosphate of lime, and have, as was before indicated, nothing to do with gouty deposits. Besides the above mentioned symptoms, they give rise to frequent angina and hypertrophy of the tonsils.

The diagnosis can only be made, when pieces of the structures come away, or when they can be seen and felt through the tissues.

Treatment.—The treatment must be directed towards filling up the depressions and openings in which the softer concretions and pieces of food collect. In some cases I was able, by breaking down the bridge-like bands, and by exciting adhesive inflammation and obliteration by applying the galvanic cautery, permanently to remove the symptoms, and in others by amputating the fissured tonsils. Larger tonsil stones must be enucleated by means of forceps or spoon-shaped instruments; if the tonsil is hypertrophied, amputation is recommended, which here is best accomplished with a bistoury, as with it one can get round the stone, while with the tonsillotome it cannot be done without danger of destroying the circular knife.

NEOPLASMS AND TUMOURS.

Neoplasms of the throat, with regard to their frequency, take a subordinate position.

By many authorities, adenoid vegetations are regarded as neoplasms of the naso-pharynx. We, however, have regarded them as forms of hypertrophy in a normal constituent part of the body, and have therefore discussed them in another place.

Polypi of the naso-pharynx will be considered along with neoplasms of the nose.

Retro-pharyngeal sarcomata, which perforate the base of the skull, and extend into the cerebrum and cavities of the nose, or cerebral tumours, which force their way through the sphenoid bone into the pharynx, are repeatedly described. Among the tumours of the naso-pharynx, fibro-mucous polypi, enchondromata, sarcomata, and medullary sarcomata, are yet to be mentioned.

The soft palate is a favourite situation of neoplasms. Mucous cysts, fibromata, lipomata, papillomata, and angiomata, especially those on the uvula and pillars of the fauces, are among the tumours occurring most frequently, while, according to Froelich, myxomata and adenomata occur very seldom indeed. The latter, called by Fonnegra encapsulated epitheliomatous glands, are always situated on one side of the middle line of the anterior surface of the soft palate; they are developed from the submucous glands, and do not become adherent in any way to adjacent tissues. Their outer surface is smooth or slightly tuberculated, their consistence compact and elastic, their size varies from that of a hazel nut to that of a hen's egg, they have a fibrous capsule, which renders their enucleation very easy.

The tonsils too are often attacked, but, as it appears, more by malignant neoplasms, such as sarcomata, cancroids, and epitheliomata; nevertheless all kinds of benign growths also occur, and amongst them fibromata, cysts, and echinococci.

Tumours very seldom have their seat in the deeper sections of the pharynx. The papillary tumours recorded by Luschka and Sommerbrodt, and a plexiform sarcoma of the laryngeal section of the pharynx described by Ehrendorfer, are still looked on as rarities; I myself once observed a pedunculated fibroma on the pharyngeal side of the aryepiglottic ligament, and a diffuse flat papillary cancer on the posterior wall, which encroached on the larynx, and gave rise to bilateral paralysis of the recurrent nerve.

The symptoms which neoplasms of the pharynx excite, are caused by their situation, size, and form. Smaller neoplasms on the soft palate and the tonsils are almost always without symptons, and are generally only discovered by accident. Larger neoplasms excite disorders of speech, difficulty in swallowing, and the sensation of a foreign body in the throat; and when they come in contact with the inner surface of the epiglottis, they cause coughing, and afterwards difficulty in breathing. Malignant tumours of the tonsils give rise to the same symptoms as hypertrophy; if they slough, and sometimes even before, they cause violent radiating pains in the ear; they seldom remain localised, but generally encroach upon neighbouring parts, even reaching the external surface of the neck, as well as affecting the glands. The pains and difficulties in deglutition are very serious when the neoplasms are in the laryngeal part of the pharynx; disintegrating cancers produce fetor of the breath, profuse secretion, and give rise to hæmorrhage, weakness, emaciation, and cachexia.

The diagnosis is generally easy, especially for those who are practised in rhinoscopy and laryngoscopy; but it is sometimes difficult to determine the histological structure of the tumours. Malignant neoplasms are always accompanied by swelling of the lymphatic glands.

The prognosis depends upon the situation, size, and histological structure of the growths.

Treatment.—That the treatment can only be operative, is easily understood. Small pedunculated neoplasms are removed by the cold snare, scissors, or Stoerk's guillotine, larger and more extensive ones with the galvano-caustic snare. Of course one acquainted with laryngoscopy is at a great advantage here, as well as in all manipulations. In malignant neoplasms, serious surgical operations, resection of the under jaw, or pharyngotomy, may be necessary, according to their seat and extent.

Exostoses of the spinal column and retro-pharyngeal goitre must be considered amongst the tumours of the pharynx.

Struma retro-pharyngea or accessoria, called also *retro-œsophagea*, is developed in a preformed posterior horn of the thyroid gland, which penetrates from the superior, sometimes also from the inferior pole of the lateral lobules, backwards and inwards into the so called retro-visceral space at the level of the entrance to the larynx.

As the disease develops gradually, there are no symptoms at the beginning. As time goes on, there arise difficulties in swallowing, in that harder and larger pieces of food are swallowed with more difficulty or are regurgitated, while fluids still pass easily, until at last these, too, cannot any longer be swallowed. There is generally the feeling as of a foreign body in the throat, and in one case observed by Czerny radiating pains were present in the ear. As the swelling grows, difficulties of respiration set in, at first only on movement of the body, but afterwards even when at rest. According to Kaufmann, whose description I am following, attacks of suffocation sometimes come on suddenly, due apparently to interstitial hæmorrhage and momentary increase in the size of the tumour. By pressure on the nerves of the larynx, hoarseness may be caused, by pressure on the cricoid cartilage, paralysis of the abductors; in a case recorded of Weinlechner, considerable dilatation of the lowest section of the pharynx was brought about.

On examination, the mirror shows a more or less considerable arching forward of the posterior and lateral walls of the laryngopharyngeal cavity. The swelling is round, has a complete smoothly nodular surface, the larynx is more or less pushed away from the middle line towards the side, the aryepiglottic ligament arched inwards towards the lumen of the larynx, the glottis is narrowed, as also is the trachea and the entrance to the œsophagus. Digital examination, which for the completion of the diagnosis should never be omitted, proves that the tumour is not sensitive to touch, and is of the consistence of the usual benign goitre, and that during swallowing it rises simultaneously with the swelling on the outside of the neck.

According to Kaufmann, the most certain sign is, that the tumour can be pushed towards one side, or altogether out of the throat, by the palpating finger, and made to appear externally on the neck at the level of the thyroid cartilage.

The diagnosis is founded on the above mentioned condition, and on the fact, that tumours and abscesses arising from the periosteum of the spine, or from the prevertebral connective tissue, do not, or only to a slight degree, rise up during swallowing, and can be displaced towards both sides. The diagnosis becomes certain, when the part connecting the retro-pharyngeal tumour with the lateral swelling of the thyroid gland can be felt by palpation.

The prognosis is most unfavourable; for although, for a long time, bad symptoms are absent, or insignificant, nevertheless, at a later stage, symptoms threatening life almost always appear.

Treatment.—Treatment is either palliative or radical. In any case, provided the symptoms are not too far advanced, a course of iodine should be energetically tried. Inunction may be combined with the internal administration of iodide of potassium (15 to 45 grains per diem) or the latter may be tried alone, along with the ice treatment proposed by M. Schmidt, by means of Leiter's refrigerating apparatus or of the more comfortable ice cravat. When difficulty of swallowing is very great, the patient must be fed by means of the œsophageal catheter, and when dyspnœa comes on, tracheotomy must be performed. For the radical cure of the lesion, parenchymatous injections of iodine through the mouth have been proposed, but this is not to be recommended, as reactionary swelling always occurs, so that in cases where difficulty of breathing already exists, asphyxia might come on in a few hours. Chiari records a case, in which a simple exploratory puncture gave rise to difficulty in breathing, and to an abscess which was followed by pyæmia, causing the patient's death. Since extirpation of the tumour from the mouth appears impracticable and unsatisfactory, there remains only excision from the neck. According to Kaufmann, this operation is to be performed in the following manner.

Through an incision extending from the anterior edge of the sterno-mastoid, and dividing the skin, platysma, and fascia, the omohyoid is reached, and is either partly or altogether cut through; the superior thyroid artery is now tied above and below, and cut through. The tumour is next forced out of the pharynx, the fascia still investing it along with its capsule is slit up as far as necessary, and the enucleation completed by the finger: the remaining peduncle is ligatured, and the lateral lobe of the thyroid gland extirpated.

DISEASES OF THE NERVES.

Considering the extraordinary number of nerves in the pharynx, it cannot be considered remarkable, that they very often become diseased.

We will begin, holding to the division of nerves into *sensory* and *motor*, with the neuroses of the sensory nerves.

Anæsthesia of the mucous membrane of the pharynx, may be either central or peripheral in origin. Among the central lesions, the principal are hæmorrhages, tumours of the brain, encephalic deposits, sclerosis, atrophy, and bulbar paralysis. Anæsthesia of the pharynx also occurs after epileptic attacks, in the asphyxic stage of cholera, and in hysteria; Jurasz observed it once along with motor paralysis after croupous pneumonia. But far most commonly it occurs as a sequel of diphtheria. Further, certain drugs, especially morphia, chloral, and bromide of potash, are well known to allay the sensitiveness of the mucous membrane, whether applied locally or internally. According to Fr. Seitz, anæsthesia of the pharynx sometimes occurs after epidemic catarrh (influenza).

The subjective symptoms consist sometimes in a feeling as of fur in the throat, often along with painful sensations (anæsthesia dolorosa). Generally, however, no symptoms are present.

The diagnosis depends entirely on objective examination. In unilateral anæsthesia, the affected half of the pharynx is insensible to mechanical, thermal, or chemical irritation, and as a rule the reflexes are either impaired or abolished. When bilateral, the mucous membrane is entirely and altogether insensitive. Diphtheritic anæsthesia is very often combined with unilateral or bilateral anæsthesia of the larynx.

Treatment.—The treatment depends on the cause. When due to diphtheria, and when the larynx is also paralysed either as to motion or sensation, artificial feeding by means of the pharyngeal catheter is indispensable, to guard against food entering the lungs and setting up pneumonia; unfortunately this too often miscarries on account of the obstinacy and want of reason, not only of young, but also of older patients. The local application of electricity is also recommended, in both constant and induced currents, by means of Ziemssen's double electrode. According to von Ziemssen, subcutaneous injection of strychnine ($\frac{1}{12} - \frac{1}{6}$ pro dosi) often exerts a beneficial influence, when all remedies have failed.

By *hyperæsthesia* is understood that increased sensory irritability of the pharynx, which is manifested under the influence of stimulus, which can be traced as peripheral; through irritation, increased sensation is called forth, which is either perceived as pain, or is accompanied by reflex symptoms. Hyperæsthesia of the pharynx is a very common symptom, as one may be convinced of any day, when examining with the laryngoscope; it is often so very great, that not only touching the pharynx, but even approaching an instrument towards the mouth, or simply putting out the tongue, brings on most violent coughing, retching, and spasmodic paroxysms. Besides this, local diseases of the pharynx, acute and chronic catarrhs, general hyperæsthesia, and psychical excitements, must be mentioned as etiological factors.

Treatment.—Hyperæsthesia of the pharynx is not easily accessible for efficient treatment; but some result is soonest attained by frequent touching with the probe, spraying with cold water, cold gargles, or painting with tannin or with Mandl's solution. Jurasz recommended painting with the following:—

℞ Potass. bromid.	grs. 20
vel. Chloral hydrat.	grs. 20
Aquæ destil.	℥i
Morph. mur.	gr. ½

In conclusion, I must say, that I have never succeeded in producing anæsthesia of the pharynx or larynx by the local application of ether or chloroform.

Pure *neuralgia of the pharynx* is very rare. Still I have seen several indisputable cases; they occurred, however, only in hysterical women, who, although the mucous membrane was perfectly normal, complained of violent attacks of darting lancinating pains. Distinct painful points can almost always be discovered; in several, the parts at the side of the hyoid bone, or the larynx and trachea, were sensitive to pressure, and in another, a circumscribed spot on what appeared to be a perfectly normal right tonsil.

Paræsthesia is undoubtedly the sensory disturbance which occurs most frequently in the pharynx.

By this term is understood abnormal sensations, which generally assert themselves as burning, tension, irritability, dryness, rawness, itching, or as the feeling of the presence of a foreign body. As regards the last, the most manifold differences are met with, according to the stage of development and amount of attention given to it by the patient. Sometimes it is a piece of bone, or a fish-bone, a hair or a bit of glass, or a needle, sometimes a grain of sand or dust, that is said to be sticking in the throat. The best known is the feeling of a ball rising up and down in hysterical persons (globus hystericus), or the complaint of a burning pain or feeling of icy coldness during inspiration. Most patients inspect their pharynx daily, sometimes hourly, finger their larynx, and become quite bewildered by the noise it makes when it moves during swallowing. Whether or not the perverted sensations—especially the feeling of tension—are caused, as A. H. Smith maintains, by paresis of the floor of the mouth, must remain uncertain.

Paræsthesia is a constant accompaniment of chronic catarrh of the pharynx, and it is often met with after removal of foreign bodies, or in consequence of extremely insignificant changes in the mucous membrane. Pure paræsthesia, independent of disease of the mucous membrane, most frequently occurs in hysteria and hypochondriasis, in the latter case, as I can affirm with Jurasz, generally in consequence of reading medical books, or from fear of diphtheria, or the re-appearance of former healed syphilis. Those classes are most severely attacked, whose occupation demands excessive use of the vocal organs, as also are anæmic and chlorotic persons, women who suffer from gastric or uterine disease, and persons of both sexes who are easily excited; more rarely it is caused by diseases of the brain, or by bulbar paralysis.

The course of the disease is very chronic; there is no affection which tries the patience of the patient, and of the physician, more than this.

Treatment.—The treatment of paræsthesia and neuralgia must, in the first place, be directed towards the cause of the affection. Pharyngeal catarrh, and other local affections, which may not be apparent, also anæmia and chlorosis, must be removed. One may often succeed in curing fundamentally and finally the hypochondriac and the fearful by a little sensible instruction, or by taking away the now so abundantly circulated popular medical rubbish. In general nervousness, or in hysteria, the inward administration of drugs which act upon the nerves is recommended, although they generally only produce temporary benefit, especially bromide of potassium (grs. 45–75 daily); also valerianate of zinc and arsenic. For globus hystericus Roth recommends 20 drops of tincture of pyrethrum root. I have seen good results from mild hydropathic treatment—rubbing, shampooing, lukewarm baths, associated with the cold shower, douches, salt or seawater baths. In every case a trial of galvanism is indicated; the positive pole is laid on the soft palate, or on the posterior wall, the negative on the skin of the neck near the hyoid bone, or on the painful spot; others recommend the application of the electrodes in the opposite way. The internal or subcutaneous administration of narcotics is to be particularly avoided, as in nearly all cases they do no good, and, moreover, there is always the danger of converting the patient into a morphinist.

To be regarded as a *vasomotor neurosis* is the reddening and injection of the mucous membrane of the pharynx, first observed by Rossbach; it comes on suddenly, lasts from several minutes to half an hour, and then goes off again, leaving the mucous membrane normal in colour. The persons thus affected suffered from pronounced hyperæsthesia of the œsophagus; and although no immediate connection of the latter with the injection of the mucous membrane could be proved, still Rossbach believed that to some extent a connection existed between the two conditions, as both the sensory nerves of the throat and their centres, were (in the same manner as the vasomotor), unusually easily excited.

Disturbances of the motor nerves occur almost as frequently as those of the sensory.

They are thus divided: increased mobility or spasm, and diminished mobility or paralysis.

Spasm of the muscles of the soft palate, of the constrictors, and of the first part of the œsophagus, are disorders which seldom occur. As already remarked, they sometimes occur reflexly in the course of chronic pharyngeal catarrh; in tetanus and hydrophobia they are never absent, and more rarely they occur as consequences of brain diseases, particularly of irritable conditions, and of hyperæmia of the brain and its membranes. They are most frequently tonic, especially in hysterical persons (dysphagia hysterica); clonic spasms of the pharynx are unusually rare, sometimes they attack individual muscles of the soft palate, as the communications of Böck, Moos, Politzer, and Schwartze have shown, sometimes all the muscles of the palate, as occurred in a case observed by Küpper. A rhythmical rising of the larynx, of the floor of the mouth, and of the soft palate, was seen in this patient, while, at the same time, the pillars of the fauces approached one another; during the spasm a peculiar sound could be heard, not unlike that produced by rubbing the finger nails against each other.

Paralyses of the pharyngeal muscles are clinically of most importance.

They are usually of central origin; acute and chronic diseases of the brain, especially those which lead to compression of the vagus, spinal accessory and glosso-pharyngeal nerves, also degenerative atrophic processes of the brain, such as bulbar paralysis, locomotor ataxia, and progressive muscular atrophy, are the most frequent causes. Facial paralysis, too, is in some cases followed by motor disturbances of the soft palate. Among the peripheral causes, diphtheria takes the first place, then come local diseases, which give rise to functional disturbances by serous infiltration, and inflammation of the muscles, and must be separated from paralyses proper. Sensory and motor paralysis very often occur simultaneously.

Paralysis of the soft palate occurs most frequently; it may be unilateral or bilateral, partial or complete. When unilateral, the soft palate and uvula are drawn towards the non-paralysed side, the arch of the fauces on the affected side is abnormally large and wide, that of the sound side being narrow and small; during phonation there is distinct distortion towards the healthy side. In paralysis of the uvula, a symptom frequently accompanying acute pharyngeal and laryngeal catarrh and paralysis of the vocal cords, it in like manner inclines to the sound side.

In bilateral paralysis of the soft palate, the uvula hangs down loosely, and shows no trace of active movement, but during respiration it is moved backwards and forwards like a loose curtain. During phonation in partial paralysis, a slight movement upwards of the palate can be noticed; when the paralysis is complete, it remains absolutely immoveable.

The diagnosis of paralysis of individual muscles of the soft palate, first attempted by Duchenne, is still, even now, hardly possible.

Paralysis of the constrictors, with or without participation of the soft palate, and of the œsophagus, is generally a sequel of diphtheria, or of bulbar paralysis.

If the paralysis is limited to the constrictors, it first manifests itself during the act of swallowing. Hard pieces of food pass over the dorsum of the tongue, but stick at its root and in the epiglottic fossa, and must, if they do not succeed in going further, be removed with the fingers, on account of the difficulties of respiration which arise. Fluids easily fall into the larynx, and excite coughing and choking. When the paralysis is unilateral, patients are unable to swallow the food passing down the paralysed side. If the superior constrictor of the pharynx alone is paralysed, the food may be returned through the nose, since it, along with the soft palate, effects the closure of the naso-pharynx.

The disturbances taking place in complete bilateral paralysis of the soft palate are very remarkable. Speech is nasal, some words can hardly be understood, and the indistinctness reaches its highest degree, when, as occurs in bulbar paralysis, the tongue and lip are also paralysed. As it is impossible to shut off the naso-pharynx from the mouth, it is easy to understand how the food comes to be ejected through the nose.

Treatment.—The treatment of paralysis of the œsophagus must be mainly electrical. Although many paralyses, such as those following diphtheria, resolve of themselves, still their cure may be accelerated by means of electricity. The constant current is preferable to the interrupted, the positive pole is placed on the cervical vertebræ, while the negative is moved along the neck close under the angle of the lower jaw, or still better, is placed directly on the paralysed muscles inside the pharynx. The already mentioned subcutaneous injections of strychnine are also of great benefit in these conditions.

SECTION III.

DISEASES OF THE NOSE.

DISEASES OF THE NOSE.

ANATOMICAL AND CLINICAL INTRODUCTORY REMARKS.

THE nose may be considered as a cavity, widely open both in front and behind, and divided into two halves by a thin partition or septum. The posterior division is on all sides surrounded by bone, while the anterior, or the external nose, is made up of bones and of non-vascular cartilages.

The external nose has as its framework several hyaline cartilages, which are united to one another, and to the pyriform aperture, by strong connective tissue. Connected with the cartilage of the septum, which occupies the quadrangular space between the vomer and the perpendicular plate of the ethmoid bone, are the triangular cartilages. According to Rüdinger, these cartilages go to form the anterior curved lamellæ of the cartilage of the septum; inferiorly they are connected with the cartilages of the alæ nasi in such a way, that the latter form a wide arch, and can be pushed over the upper cartilages. Both cartilages of the alæ nasi are firmly bent forwards and inwards, so that the inferior end of the cartilage of the septum can be felt in the furrow between them. In the angle between the inferior alar cartilage and the triangular cartilage lie the sesamoid cartilages—several small cartilages which are embedded in dense fibrous masses. The external nose is supported on the nasal bones, two longish quadrilateral bones meeting in the middle line, which form the convex bridge and root of the nose. At their place of union above, the perpendicular plate of the ethmoid bone runs backwards, and the cartilage of the septum downwards. The frontal processes of the superior maxillary bones also take part in the formation of the nose. Their inner surfaces, as well as those of the nasal bones and their cartilages, when covered with mucous membrane, form the anterior superior wall of the nasal cavity. The external skin of the nose is characterised by the abundance of sebaceous glands in it, which form the well-known comedones when their contents become pent up. The muscles which dilate and contract the nostrils are the levator alæ nasi, the depressor alæ nasi, the levator alæ nasi proprius, and the compressor narium minor. The arteries of the external nose spring from the branches of the facial artery; the dorsal artery of the nose, the external maxillary, and branches from the ophthalmic, also supply the external nose. The veins are more numerous. They unite with the anterior facial veins, and according to Rüdinger a venous plexus lies symmetrically at the entrance of the nares, deep under the skin, into which plexus the blood of the anterior part of the nares flows. The muscles of the nose receive their nerve supply from the facial; the sensory nerves of the skin are branches from the infra-orbital, those supplying the skin of the tip of the nose spring from the nasal branch of the first division of the trigeminus.

Several bones take part in the formation of the nasal cavities.

The superior wall, or roof, is formed anteriorly by the two nasal bones, in the middle by the ethmoid, and posteriorly by the anterior wall of the sphenoidal sinus. The inferior wall, or floor of the nose, is formed by the palatine process of the superior maxilla and the horizontal plate of the palate bone. The floor of the nose is smooth, and somewhat sloping from before backwards. The inner wall consists of the septum; its bony part is formed by the vomer, the vertical plate of the ethmoid, by the crests of the palate and superior maxillary bones; its cartilaginous part by the cartilago septi, which very often exhibits changes of form, chiefly bulging and exostoses. The outer wall is the most complicated. It is formed by the superior maxilla, the palate bone, and the wing-like processes of the sphenoid. Zuckerkandl distinguishes an upper part, reaching to the insertion of the inferior turbinated bone, and a lower, reaching from the inferior turbinated bone to the floor of the nose. The upper part, behind the frontal process of the superior maxilla, is provided with a large irregular cavity, which is covered by the pterygoid process of the palate bone, the ethmoid process of the inferior turbinated bone, and the unciform process of the ethmoid bone. From thence to the base of the skull, the os planum of the ethmoid bone and the lachrymal bone form the outer wall. In the outer wall are several lacunæ, varying in number, form, and size, which are covered over with mucous membrane; the largest, a half-moon-shaped opening two centimetres long, is called the *hiatus semi-lunaris*, or ethmoidal fissure, and leads into a funnel-shaped cavity called the infundibulum, which, in its anterior and upper part, is in connection with the frontal sinus, and at its inferior and posterior part with the maxillary sinus by means of the *ostium maxillare*.

The most important formations on the lateral wall of the nose are the turbinated bones.

The largest, the inferior turbinated bone, is an independent thin bone of the face, which is attached to the inner surface of the upper jawbone, and whose inferior border is free and extends into the nasal cavity. The space which remains free between the inferior turbinated bone and the septum on the one side, and the floor of the nose upon the other, is the inferior nasal meatus, the size of which mainly depends upon the configuration of the inferior turbinated bone, and through which are introduced all instruments which are used in operating on the nose and naso-pharynx. Under the inferior turbinated bone, there is a little oval opening extending outwards and forwards, which is the efferent duct of the naso-lachrymal canal. The middle turbinated bone is a part of the ethmoid, and is placed not only higher but also further back than the inferior, and its anterior end reaches free into the nares ten or twelve millimetres. Its inferior border is generally broad and flat, and forms, with the anterior border, a more or less obtuse angle. The free space, which extends downwards from its inferior border to the inferior turbinated bone, is the middle meatus of the nose; the narrow space lying between it and the septum is called the *olfactory fissure*, and contains in its mucous membrane the endings of the olfactory nerve. The olfactory fissure is sometimes entirely closed by thickening over the middle turbinated bone, or from deviations of the septum. In the middle meatus lie the orifices of communication of the maxillary and frontal sinuses. The smallest of the turbinated bones is the superior; it is also a part of the ethmoid, is still shorter, and less twisted than the middle, and is only differentiated from the latter in its middle and posterior thirds; anteriorly the two bones coalesce; if the superior turbinated bone be divided, then there comes to be a more superior or fourth turbinated bone, which, according to Zuckerkandl, is always present in new-born children. The space between the middle and superior turbinated bones is the superior nasal meatus; into it the ethmoid cells open with numerous orifices; above and behind the superior turbinated bone is the small aperture of the sphenoidal sinus.

From a clinical point of view the mucous membrane of the nasal cavity, especially the part, the seat of the sense of smell, called the olfactory or Schneiderian membrane, is of very great interest. The mucous membrane, connected anteriorly with the skin of the face and posteriorly with the mucous membrane of the pharynx, covers the septum, turbinated bones, and all cavities uniformly, and is very closely connected with the underlying tissue, especially that of the turbinated bones. A characteristic of the normal mucous membrane of the nose, is the impossibility of making deep impressions in it with a probe, or of pushing it up in front of the probe; it is so closely connected with the periosteum, that nowhere can the two be distinctly separated from each other. The mucous membrane has at its superior part, the olfactory region, ciliated epithelium; at its lower part, the respiratory region, it is invested with pavement epithelium. The nasal mucous membrane is abundantly supplied with nerves, blood-vessels, and glands.

The nerves of the nasal cavity are partly sensory nerves and partly nerves of smell. The former spring to a large extent from the first and second branches of the trigeminus (the spheno-palatine nerve, the Vidian nerve, the naso-palatine nerve, the nasal branch of the ophthalmic nerve, and the anterior palatine nerve), and possess in an extraordinary manner, the power of accomplishing the most varied physiological and pathological reflex actions, and of transmitting them to other nervous regions.

The olfactory nerve is exclusively the nerve of smell.

It lies in a furrow on the inferior surface of the anterior lobe of the brain; anteriorly it converges towards that of the other side, and on the cribriform plate of the ethmoid bone enlarges into a somewhat oval bulb, the bulbus olfactorious, from the under surface of which proceed two rows of thin white fibres, which enter the nasal cavity through the apertures in the cribriform plate. Here they form networks, which stretch downwards on to the septum and the inner surfaces of the two upper turbinated bones, and send brush-like fibres arranged in groups into the mucous membrane. These fibres change into the olfactory cells, which consist of oval cell bodies with distinct nuclei, and terminate in fine rhabdoidal processes.

Blood vessels are also very numerous in the nasal mucous membrane. The richer in blood a part is, the thicker it appears. While the mucous membrane in the olfactory region is thin, it attains a very considerable thickness on the inferior and middle turbinated bones, and especially on the anterior and posterior ends of the inferior turbinated bone. Here the bone is covered with *cavernous spongy tissue*, the trabecular framework of which consists of parallel connective tissue fibres with distinct contours, with connective tissue corpuscles, and numerous elastic and muscular fibres. Within the trabecular framework lie very numerous vascular spaces, lacunæ and veins, which penetrate the bones by numerous small apertures. According to Zuckerkandl, the function of the spongy tissue is to regulate the size of the nasal cavity, and to prevent any abnormal dilatation of it; the great abundance of blood-vessels in the spongy reticulum, and the compactness of their situation, further serves to keep the mucous membrane warm as well as moist. The blood of the spongy tissue flows, on the periosteal side of the mucous membrane, through numerous canals in different directions towards the face, the palate, the pterygo-palatine canal, and the naso-lachrymal canal. Of great clinical importance is the fact that the spongy tissue may suddenly increase and decrease in volume, on account of the most different kinds of irritation, even in distant organs; very often there suddenly appear red nodular swellings obstructing the nose, which after a while disappear and leave no trace. The increase and decrease of swelling in the cavernous tissue is produced through the facial nerve, and specially the spheno-palatine ganglion.

The veins of the nasal cavity accompany the arteries; they are, however, more numerous and larger, and communicate with the facial and ophthalmic veins; some of them are transmitted through the ethmoid plate and the foramen cœcum into the longitudinal and coronary sinuses.

The arteries reach the nasal mucous membrane from several

sources. The spheno-palatine artery, with its post-nasal branches and the branch to the posterior part of the septum, springs from the pterygo-palatine artery; the anterior and posterior ethmoidal arteries from the ophthalmic, and the artery to the anterior and posterior parts of the septum from the external maxillary. They form numerous anastomoses, and pass into the spongy tissue on the inferior turbinated bone, in which, however, the veins predominate.

The lymphatic vessels of the nose unite, according to E. Simon, into a common trunk, which runs between the anterior end of the Eustachian tube and the posterior end of the turbinated bone. Here they form a network, from which two or three branches spring, these extend backwards and outwards between the levator and tensor veli palati, whence the one branch passes along the outer wall of the pharynx, between the internal carotid and the stylo-pharyngeus muscle, into one of the lymphatic glands lying in front of the spinal column; the other penetrates the digastric, passes outwards from the lingual nerve, and divides into two branches, which open into two lymphatic glands situated below the sterno-mastoid.

The number of glands scattered throughout the nasal mucous membrane is very great. In the olfactory region they are long and tubular, with lateral expansions, others again are acinous; they are largest and most numerous on the turbinated bones, smaller and less abundant on the septum.

In close communication with the nares are the *accessory cavities, sinuses,* or *pneumatic spaces.*

The largest of these is the *maxillary sinus,* or *antrum of Highmore.* Its shape is that of an irregular pyramid, its apex being towards the zygomatic process; its lateral walls are formed by the orbital cavities and the lateral plates of the superior maxillary bones; its base, or inner wall, which separates it from the nasal cavity, consists of portions of the superior maxillary, palate and inferior turbinated bones, and the unciform process of the ethmoid bone. It communicates with the nose by a round or slit-shaped opening, the lumen of which varies very much. The opening lies at the level of the anterior end of the middle turbinated bone, in the middle meatus; there is often a second aperture of communication below the centre of the middle turbinated bone. According to Reschreiter, the antrum of Highmore in men, always reaches a lower level than the nasal cavity. The lining membrane of the antrum contains acinous and tubular glands, and serves partly as mucous membrane, partly as the matrix of a periosteum for the walls of the cavity.

The *frontal sinus* resembles a three-sided pyramid, whose apex is in the ascending part of the frontal bone, and whose base is formed by the upper surface of the orbital plate. Its form and size is very variable; it often extends only very slightly beyond the level of the superciliary ridge, and in a horizontal direction reaches only as far as the anterior section of the orbital roof; sometimes the frontal sinus extends out as far as the zygomatic process of the frontal bone, reaches far up into the squamous portion, and communicates with the large pneumatic spaces which occupy the orbital roof. In many persons the frontal sinuses are very small and scarcely to be distinguished, in others altogether absent, and often they are very unequally developed in one and the same person; they communicate with the nose by an opening, which is situated under the anterior end of the middle turbinated bone.

The *sphenoidal sinus* has a very obscure situation. Its anterior, upper, and lateral walls are thin and compact, while on the posterior and inferior wall there is a layer of spongy osseous tissue; sometimes the sinus is separated from the cranial cavity by a thick lamella of bone. Sometimes it is altogether wanting, while in other cases it is unusually large, and sends expansions into the basilar process of the occipital bone, and into the large and small wings of the sphenoid bone. Its aperture of communication with the nose is situated above and behind the superior turbinated bone.

The *air cells of the ethmoid* or *ethmoidal sinuses* are of subordinate importance.

They are situated in the lateral masses of the ethmoid, and consist of cells of very varied size and number, which communicate with one another and also with the nasal cavity, and are divided into the anterior, middle, and posterior cells. The lateral cells are often so markedly developed as to arch forward the orbital roof. The middle and posterior cells open into the superior nasal meatus, the anterior open towards the concave surface of the middle turbinated bone.

The physiological significance of these accessory cavities has not yet been explained. While some believe that they are there only to lessen the weight of the skull, others consider that their function is to secrete a mucous fluid, by means of which the nose is kept moist.

Braune and Clasen look upon the former opinion as improbable, because the skull, when these cavities are filled with osseous tissue instead of air, proved to be only a trifle heavier, scarcely worth considering. Against the other opinion may be placed, not only the nature of the mucous membrane of the sinuses, which is very thin, poor in vessels, and unsuitable for a regular production of mucus, but also the high situations of the openings, out of which the secretion could only flow when the body was in an unusual position. They believe, therefore, that these cavities serve for the better perception of smell. They found further, that on inspiration there was a rarefying of air in the accessory cavities, on expiration, a condensation of the air.

Paulsen also obtained similar results. He found that under ordinary circumstances, when a continuous current of air was passing through, the exchange of air in these cavities was a most insignificant one, but that it at once becomes more plentiful when the current of air is from time to time interrupted, or, as under normal circumstances, is directed at one time inwards, at another outwards. Kessel has a peculiar idea. He says that the nose and its accessory cavities, along with the middle ear and the cells connected with it, form a system of spaces lying round about the anterior and middle cerebral lobes, which supply the latter with vapour, carbonic acid, ammonia, and warmth, and precisely on account of these products, may be looked upon as the respiratory organs of the anterior and middle lobes.

METHODS OF EXAMINATION.

In the examination of the nose two methods are employed: inspection from in front, or anterior rhinoscopy; and inspection from behind, or posterior rhinoscopy. Although the latter method may furnish important details concerning the naso-pharynx and the posterior part of the nasal cavities, yet to the former method, *i e.*, anterior rhinoscopy, is imputed the greater significance. But to make the examination complete, both methods should be employed in every case.

For anterior rhinoscopy certain instruments are necessary; a nasal speculum (Fig. 4), whatever be its form or construction, is used to dilate the nostril, and thus allow the entrance of as much light as possible. As practice and custom influence so much the choice of a speculum, it makes it all the more difficult to say which of the many varieties deserves the preference, whether that of Jurasz, or that of B. Fränkel, or those of Duplay, Bresgen, Voltolini, and Zaufal. I myself still use the instrument I had when I first studied rhinoscopy; it is a Kramer's ear speculum with two blades, which open out when pressure is made on the handles, and close again by a spring. I consider it a great advantage that the position of the branches can be modified, by increasing or lessening the pressure at any moment, without the assistance of the other hand, a comfort also felt by the patient, particularly when the operation lasts for some time. After the light has been concentrated by the forehead mirror, in the same way as in laryngoscopy, the handle of the speculum is held in the palm of the left hand, so that the dorsal surface of the fingers is turned towards the patient's face. The speculum, having been warmed over the lamp in cold weather, is now introduced into the nostril to be examined one or two centimetres, and the branches separated from one another. The hairs which sometimes interfere with the view are, if necessary, pushed to the side by slightly rotating the speculum. It is recommended not to introduce the point of the speculum further than the anterior end of the inferior turbinated bone, in order that morbid changes of the same may not be concealed. With the right hand now laid on the top of the patient's head, it is necessary for the examination of the direction of the inferior turbinated bone, as well as of the inferior meatus, the floor of the nose, and the lower part of the septum, to bend the patient's head a little forwards and downwards; for the examination of the middle turbinated bone, the middle meatus, and the upper part of the septum, the head must be horizontal; and for that of the nasal roof, backwards and upwards; while at the same time, the axis of the speculum is made to follow these movements of the head. To allow the patient to choose the position

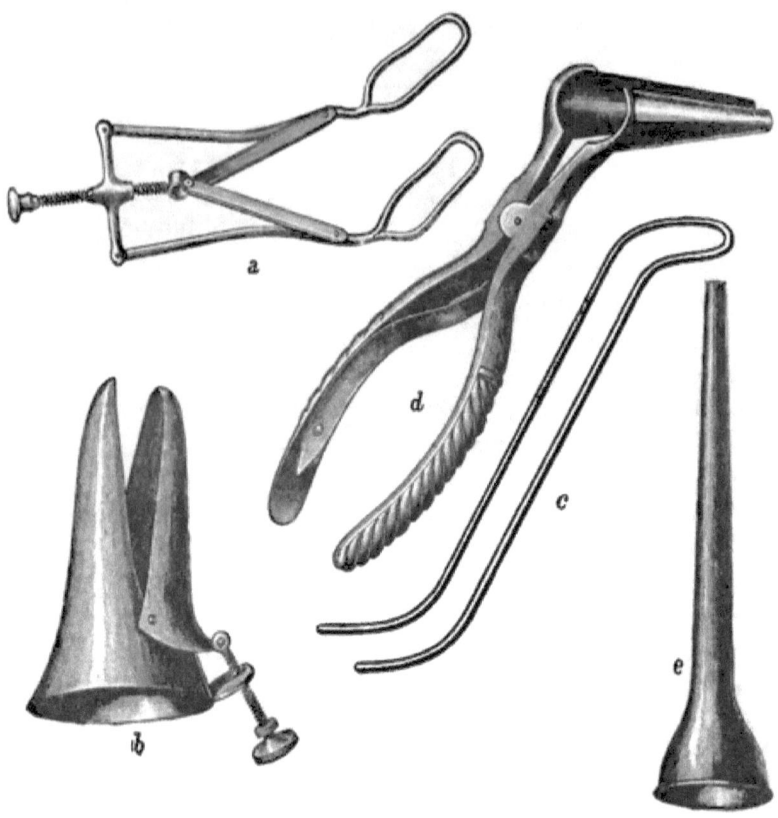

Fig. 4.—Various Nasal Specula.

(*a*) B. Fränkel's. (*c*) Jurasz'. (*e*) Zaufal's Naso-pharyngeal
(*b*) Duplay-Charriere's. (*d*) Kramer's. speculum.

of the head is useless and a waste of time; any collected secretion or blood must first be removed, by blowing the nose or by syringing. It is often impossible to avoid causing slight hæmorrhage, even in the most careful examination.

The nasal septum appears as a reddish yellow, more or less straight, vertically situated plate; the inferior turbinated bone, as a hemispherical, sometimes flask-shaped reddish prominence; and the floor of the nose as a flat or slightly hollow furrow. The middle turbinated bone is situated higher and further back, and nearer the septum, so that one is not often in a position to follow its inner surface very far back. The inferior meatus is normally the widest, the middle less wide; the superior turbinated bone and the superior meatus cannot be seen from in front.

The numerous varieties of internal contour met with in the nose are discussed under "Malformations and other Anomalies of the Nose." For the present, it is enough to mention that it is often most difficult for the beginner to make out properly what is seen. How far it is possible to see into the nose depends on each individual case; straight and wide noses often allow the whole interior to be seen, while, when the space is encroached on, it is often only possible to see the anterior end of the middle turbinated bone, or perhaps not even so much.

That the examination by means of Zaufal's speculum, which was previously described, may also furnish valuable conclusions, does not require to be specially noted.

Palpation by means of a probe is quite indispensable for an exact examination of the nose; it may not only reveal to us whether the inspected parts are normal or diseased, but also helps to determine the condition of the mucous membrane, and the manner in which all kinds of neoplasms are attached to it. It is also useful for diagnosing foreign bodies, concretions, and affections of the bones, and for pushing aside any portions of the mucous membrane which are swollen and obstruct the view. For this last purpose the nasal spatula invented by Zaufal is much to be preferred; it is a broad, plane-concave, hollow probe, and with it the hypertrophied and relaxed mucous membrane may be compressed and pushed aside. The application of this instrument, of which a special one is required for each nostril, is less uncomfortable for the patient, than that of specula with very long blades, which are introduced far into the nose; besides, it is often of great use in operating.

For the diagnosis of nasal affections, the senses of smell and hearing are also brought into requisition. The former gives us information during expiration through the nose, whether or not the nose contains products of decomposition, and the latter instructs us as to the condition of the voice, nasal timbre, etc., as well as to the permeability of the nose, in that it recognises whether the respiratory murmur is changed (as in stenosis) or altogether awanting.

The formerly practised percussion of the nose has now become unnecessary, owing to the modern methods of examination.

GENERAL THERAPEUTICS.

In treating all cases of disease of the nose, cleansing is the first essential.

In poor practice, and also otherwise under special circumstances, the snuffing up of fluid is the simplest method, but at the same time the least efficient. The patient is allowed, either directly to draw fluid up into the nose from a glass ("drink through the nose"), or to draw it up from the hollow of the hand, while the head is bent backwards. For awkward people, or for children, a spoon or a small beak-shaped vessel, the so-called "nose bath," is best suited for introducing the fluid into the nose. Politzer uses the nasal bath for the application also of various remedies. Puricelli allows the nasopharynx to be filled while the patient says "ah;" von Tröltsch and Mosler recommend that the gargle be thrown out through the nasal fossæ.

Apparatus specially designed for cleansing the nose, are called nasal douches.

Weber's nasal douche is by far the most commonly used. It is simply an indiarubber tube, 1 to 2 metres long, provided at one end with a perforated metal plate, which is placed at the bottom of the vessel containing the fluid to be employed, and at the other with a conical or olive shaped nozzle, which is introduced into the nose; and as the apparatus is on the syphon principle, the fluid must first be drawn through the tube.

The irrigator is a very practical apparatus, because it can be put to other uses (enema, uterine douche, and stomach tube), and consists of a tin vessel of any size, which is hung up on the wall, the outflow tube of which may be furnished with a stop-cock or not as desired.

In order to propel the fluid with greater force into the nose, and to remove the often extremely tough secretion, syringing is required, either with a piston syringe, or one acting by suction and pressure, the latter being known as the English, or Davidson's enema syringe. It consists of a balloon with a valve apparatus, which has at each of its conical ends a tube ¼ to ½ a metre in length; one end weighted with lead is placed in a vessel, the bone nozzle of the other tube, which is the longer, is pushed into the nostril, brought into a horizontal direction, and laid gently against the septum. After the air is removed by pressure on the balloon and water is sucked in, the fluid is driven into the nose by slow and regulated pressure.

Other instruments used for cleansing the nose, are the steam inhaler and the cold spray with larger apertures than usual, particularly when the nose is completely impermeable, or in paralysis, defects and perforations of the soft palate.

With every kind of nasal douche, the soft palate contracts from the stimulus of the fluid pouring in on it, and thus the naso-pharynx is shut off from the mouth, and fluid syringed in flows out by the other nostril. If the douche is too long applied, the palate relaxes, and a part of the injected water gets into the mouth and pharynx, or is swallowed, an event which must be carefully avoided with some remedies, particularly the much used chlorate of potash.

As regards the temperature of the fluid, there is great diversity of opinion. While some recommend fluids very warm, at blood heat, or lukewarm, others use cool or even quite cold water. While too hot water favours congestion of the nose, quite cold water causes an extremely unpleasant burning and painful sensation, hence a middle course should be taken, and water at $77°$ F. should be begun with, and gradually cooled down to lower temperatures. In this way the mucous membrane not only becomes contracted but gradually hardened. It is self-evident that the higher temperatures should principally be applied in winter, the lower in summer.

The quantity of fluid to be used depends on the nature and amount of the secretion. Fluid or semi-fluid exudation can be more quickly removed than thick or hardened dry secretion. In the former case therefore from ½ to 1 litre will be required, while in the latter several litres are usually necessary. Enormous quantities, 20 litres, and even a whole bucketful, have been used without harm, but also without effect. As a rule, one to three applications are enough each day.

But more important than quantity and temperature is the quality of the injected fluid.

I entirely agree with Bresgen, Gottstein, and others, and use the nasal douche only as a palliative, that is to say, for cleansing, and not as an actual remedy. On this ground therefore I only order solvents, either natural alkaline and saline waters, or 1 to 2 per cent. solutions of carbonate of potash, bicarbonate of soda, chloride of sodium, muriate of ammonium and sulphate of soda; in diphtheria, diluted lime water, and also disinfectants, or rather deodorisers. First stands chlorate of potash in 1 to 3 per cent. solution; also permanganate of potash, grs. 1–2½ to ʒi of water; ½ to 1 per cent. carbolic acid, 2 to 4 per cent. boracic acid, 2 to 4 per cent. benzoate and salicylate of soda solutions, and salicylic acid or thymol, gr. ½ to ʒi of water.

From astringents such as tannin, alum, sulphate of zinc, and nitrate of silver, I have seen little good and often harm result; apart from the loss of smell they have several times been observed to cause (especially zinc sulphate, even when diluted to 1 in 1000), irritation of the mucous membrane, burning and lachrymation, and violent sneezing and headache, have been excited.

If patients find cleansing of the nose painful with harmless remedies, and complain particularly of headache, one may be quite sure that they use the douche in a clumsy and wrong manner. The error consists, apart from the temperature being too high or too low, in driving the fluid too strongly into the nose, so that the very sensitive branches of the trigeminus become irritated.

With Weber's douche, and also with the irrigator, the height of the stream of water should not at first exceed 5 centimetres. When using the syringe, too, the pressure used at first must be very slight, and then regularly and gradually increased. An even commoner mistake is, that most patients, when not specially instructed by the physician, direct the stream in a vertical direction, and therefore against the nasal roof, instead of in a horizontal direction, through the inferior meatus. Others fail, in that they do not bend the head forwards, as is the rule in using the douche, but bend it backwards. After douching, the patient should not blow the nose, but holding one nostril shut should blow out the remaining fluid through the other.

The worst reflection that can be made on the douche, is the entrance of fluid into the middle ear. Although it is true that purulent inflammation of the middle ear may be excited by unskilful manipulation, more usually, however, from closure of the nostril from which the fluid is flowing, or from swallowing movements being made at the wrong time, yet happily these unlucky cases, as I can affirm, after many years of practice, occur very rarely; but to put aside the nasal douche on account of such individual disadvantages, would be just as foolish as relinquishing the use of chloroform or morphia. Whether or not the double perforated nozzle, constructed by Berthold, can obviate that danger, remains to be seen.

As real therapeutic methods, painting, plugging with medicated wool, and insufflation, must be considered. The first is accomplished with a small brush or piece of wadding, or a piece of fine sponge; chiefly used are solutions of nitrate of silver or Mandl's solution; when, however, the space is limited, only a subordinate significance is ascribed to it. For the wool tampons which are retained in the nose for a longer time, fats, almond oil, vaseline, boracic ointment, also liquor ferri perchloridi, alum, and tannin are generally used. In a far more efficient manner, however, the nasal mucous membrane may be influenced by the insufflation of powders. The choice of these will be discussed under the different diseases. Snuffs are used partly for stimulating the mucous membrane and increasing the secretion, partly for their direct curative effect. Other remedies are used in an elastic fluid form, the vapours of which, or their fumes, are inhaled through the nose from a bottle, or from a cloth on which they have been sprinkled. The application of gelatine bougies has no special advantage.

Cauterisation of the nasal cavity is accomplished with lunar caustic or chloride of zinc; according to Hering, chromic acid is best, its crystals being melted on to the end of a suitably bent probe, and applied to the proper spot, with the assistance of the mirror. Chromic acid crystals must be cautiously heated over the flame, as they decompose if the heat is too great. Before applying caustics, syringing with luke-warm water is recommended; after the application, a soda solution (8 to 1000). That chromic acid, which may also be applied in solution on a glass rod, is able to act as effectively as the galvano-caustic point, or altogether supplant it, appears to me improbable. Hering recommended it especially in chronic swelling of the ends of the turbinated bones, for adenoid vegetations and stumps of polypi, Bayer in hypertrophy of the pharyngeal and palatine tonsils.

The most indispensable, and often almost the only remedy for successful treatment of nasal affections, is the galvano-caustic point. The instrumentarium has already been discussed, the special indications will be mentioned under the individual diseases, as well as instruments for certain purposes, such as forceps, curettes, and snares. Although lately perhaps the galvano-caustic point has been too much used, yet, in spite of Bosworth's unfavourable judgment on it, it will scarcely again be excluded from the therapeutics of nasal diseases.

Electricity only comes into use in neuroses, and then only to a very limited extent.

DEFORMITIES, ANOMALIES AND MALFORMATIONS.

If we first consider the malformations of the external nose, we meet with on rare occasions, apart from the extremely numerous formations revolting to æsthetic feeling, reduplication of the organ. Of more frequent occurrence are congenital defects, such as absence of the nasal bones, or of the nasal process of the superior maxilla; complete absence of the nose was observed by Maisonneuve. On the site of the organ was a flat surface of skin, at the lower part of which were two extremely small apertures representing the nostrils. Fissures too are pretty often seen, which generally occur in the nostrils, along with cleft palate and harelip. The anterior nares have repeatedly been found altogether or partly closed at birth; bony partitions obliterating them, according to Delstanche, who observed one such case, are very rare. The adhesions of the nostrils are generally membranous, their depth is very different, and can be determined by pressure from within, by attempting forced expiration. Acquired contractions and adhesions of the nasal apertures are very common. I myself saw a very marked case of stenosis of the left nostril from a sabre cut, and another after a burn, but most usually they occur in consequence of lupus or syphilis, from destruction of the lateral cartilages.

Mention has already been made of congenital, osseous, and hereditary membranous closure of the posterior nares, and of other anomalies of the naso-pharynx.

In the interior of the nose, congenital defects of the entire septum, of the vomer, of the inferior turbinated bone, and of the ethmoid labyrinth, are met with. Among the acquired malformations, the most common are perforations and defects of the septum, with sinking in of the external nose (Sattelnase, Nez de mouton), in consequence of syphilis and other ulcerations, while loss of the turbinated bones and destruction of the vomer are also met with. There are also very often deformities of the turbinated bones, which appear too large or too small, too flat or too much curved, grooved or even perforated. The occasional presence of a fourth turbinated bone has already been mentioned. A case of double septum has been described by Lefferts.

Among all the anomalies in form of the interior of the nose, none occur so frequently as *deviation of the septum*. It can even be affirmed that there is scarcely a single person who has an exactly straight septum; according to Mackenzie's statistics, 77 per cent. of skulls examined by him exhibited deviation of the bony septum. The curvature of the septum affects both its height and length, or only individual parts. In the most marked cases, the deviation is S shaped, and affects both nares; generally, however, only one side is affected, and that, in the majority of cases, the left. With all deviations of the septum there is a disproportion in the width of the nares; the nostril which contains the convexly arched part of the septum is narrowed, the other widened. As long as this disproportion is insignificant, there are usually no symptoms; when the deviation is very marked, there comes to be considerable disturbance of function, obstruction of nasal respiration, diminution of the sense of smell, difficulty in removing collected secretion; speech becomes nasal, while introduction of the Eustachian catheter, examination of, and operations in the nose, become excessively difficult, sometimes even impossible.

As causes of deviations of the septum, some suggest unequal growth of the skull and bones of the face, and unequal development of the cerebral hemispheres; others, with Ziem, think them due to injuries, falls on the nose on first attempting to walk, blows with fists and fractures; while Welcker considers sleeping on one side, as the cause. Ziem has proved that every obstruction of the nose exerts widely spread consequences on the development of the skull; in young animals, one of whose nostrils he completely closed up for a long time, there was seen a deviation of the intermaxillary bone and of the sagittal suture towards the shut up side; also lesser length of the nasal bone, of the frontal bone, and of the horizontal plate of the palate bone, less steep elevations of the alveolar processes, smaller distance between the anterior surface of the bony auditory capsule and the alveolar process, also between the zygomatic arch and the supra-orbital border, and smaller size and asymmetrical position of the vascular and nerve canals on the closed side of the nose. The distance of the two orbits from the middle line was unequal, which, as has also been observed in men, leads to asthenopia, astigmatism, and strabismus. Lastly, certain curvatures of the spine also seem to be due to obstruction of the nose. It is, therefore, extremely probable that deviation of the septum is the cause and not the consequence of unequal development of the skull.

Other changes besides deviation occur in the septum. The most frequent are partial thickenings and excrescences, round or spherical, and often pointed processes, also called *spines*. They are usually situated in the most anterior part of the septum, and are partly osseous, partly cartilaginous in structure. I have often seen a triangular crest occupying the whole length of the septum, at an elevation of from half to one centimetre above the floor of the nose, which, by touching the inferior turbinated bone, in a manner bisected and narrowed the inferior meatus. Michel states that a pale moveable prominence is often found near the floor of the nose, being the most inferior part of the cartilaginous septum, which has become dislocated from its connection with the osseous part. Adhesions and synechiæ of the septum with the turbinated bones, or of the turbinated bones themselves, affecting sometimes the inferior, sometimes the middle bone, not infrequently occur, and are partly membranous, partly osseous in structure; the inferior turbinated bone also is often connected with the floor of the nose by an osseous ridge.

Treatment.—The treatment of congenital or acquired malformation of the nose can, of course, only be operative. In membranous adhesion of the nostrils, the adherent parts must be separated, and their re-adhesion prevented by inserting dilating bougies. In incomplete stenosis good results are got by methodical dilatation with sponge tents or laminaria. The methods of removing osseous and membranous closure of the nasal fossæ have already been mentioned. Plastic operations for replacing the point or the alæ of the nose, and for closing clefts and fissures, belong to the province of surgery; here, however, we must turn more particularly to the treatment of deviations of the septum.

Almost the only method that has for a long time been practised—the taking away a piece of the septum, with a pair of round forceps, like those used for punching railway tickets,—has this disadvantage, that in certain conditions of the cavity, one does not succeed in getting the instrument to the level of the protuberance. As it generally happens that a piece of the part lying in front of the curvature only is cut out, the operation is really based on an illusion, for although no doubt air passes into the nostril on the curved side, yet the curvature itself still remains, so that the part of the nose lying behind it, is still without function.

Hartmann and Petersen proposed, therefore, the partial resection of the nasal septum. The patient being narcotised, the mucous membrane is split, and with a scraper is raised and pushed back from the bone beneath it. The most prominent piece of the septum, about the size of sixpence, is now cut out with a small bone pliers. A thick indiarubber tube is then inserted into the nostril, in order that the septum may be brought into the proper position and kept there.

Jurasz recommended, as Adams has done, that the bent part be pressed straight between two flat surfaces, and for this purpose he constructed an instrument, consisting of two blades which were introduced into the nose, one into each nostril, and placed against the bent septum. The instrument was next closed, and the branches so closely pressed together that the curvature was straightened and the septum forced into the middle line. This being accomplished, the blades are screwed together *in situ*, and the handles of the compressor taken away. The compressor, which is so constructed as not to interfere with the patient's eating, is left on three days and then removed, by first unscrewing and then taking the blades away one after the other.

Another method is practised by Steele. He first breaks down the protruding part of the septum by means of a forceps having four stellate sharp teeth, then inserts ivory bolts, which must be retained for four days in the proper position.

On account of the difficulty of resecting the septum, some recommend that it be left alone, and that space be provided by removal of the turbinated bones. Others, following Demarquay, seek to gain better access to the septum, by splitting up the external nose, and pushing aside the lateral cartilages and the alæ of the nose, after which a piece of the septum is taken away.

The treatment of other anomalies, which also cause more or less pronounced disturbances of function, are very various. The simplest to remove are membranous adhesions of the turbinated bones to the septum, by means of a knife or the galvanic cautery, and their contraction prevented by applying tampons. Osseous spines and synechiæ are best removed with a fine saw, or cut away with a sharp chisel. Cartilaginous and thin bony outgrowths and synechiæ can be cut away with scissors or with bone pliers, or can be destroyed or at least diminished by means of the galvanic cautery.

ACUTE CATARRHAL RHINITIS.

Synonym.—Rhinitis catarrhalis acuta.

Of all mucous membranes, that of the nose exhibits the most marked disposition to catarrhal inflammation. The name *catarrh* is given from the most prominent symptom,—the flowing down of the secretion,—and is used only with reference to acute cold in the head in many districts, though principally in South Germany.

The etiology of acute coryza is very extensive. The most important causes are undoubtedly the influence of the atmosphere, rapid change of temperature, chills, wetting the skin of the head, neck, trunk or feet, and draughts. Acute rhinitis occurs to a great extent, during a change from warmer to colder weather, or *vice versa*, and also in spring and autumn as influenza. That influenza is contagious is without doubt, but whether the micro-organisms discovered by Seifert in the secretions of patients suffering from influenza are really the cause or not, requires still to be confirmed. Acute rhinitis is a constant accompaniment of measles, often also of smallpox, scarlatina, and typhoid fever. It very often spreads from the pharynx or larynx, or is the result of irritation, which directly attacks the nasal mucous membrane. Such irritations are hot vapours, very cold air, organic or inorganic dusty substances, flour, dust, soot, sawdust, stone dust, ipecacuanha, and the pollen of different grasses. The form excited by the last named is called hay catarrh, or hay fever, about which we shall speak more particularly later on. Amongst chemical substances acting as specific irritants are, chlorine and ammonia vapour, also iodine, mercury, bichromate of potash, arsenic, osmic acid, and among others also digitalis. That persons who snuff are very seldom attacked by catarrh, is explained by the blunting of the mucous membrane and consequent diminished pre-disposition.

Catarrh occurring sporadically is also looked upon as contagious, although as yet it has never been produced experimentally. The gradual infection of all members of a family, as well as those persons who have never had catarrh, after intimate intercourse with those affected by it, and the simultaneous enlargement of the spleen, point with the greatest probability to the contagiousness of many forms of catarrh.

In many persons the predisposition to acute inflammation of the nose is extraordinary. Just as there are persons who after each chill get muscular or articular rheumatism, there are others who regularly acquire acute rhinitis. Of such, the same holds good that was said of acute pharyngitis as to defective hygiene of the skin.

The outbreak of the disease is often preceded for several days by giddiness, weariness, lassitude, disinclination for work, itching in the naso-pharynx, chilliness, and shivering. Along with these prodromal symptoms, there may soon appear a burning prickling sensation in the nose, violent beating pain in the forehead, sneezing, epistaxis, and increased secretion. The secretion rapidly increases, and often reaches such an intensity, that patients are obliged day and night to hold a handkerchief to the nose. At the beginning and at the height of the disease the secretion is thin, clear like water, acrid, and irritating, in consequence of its containing salt and ammonia. Smell is diminished or altogether destroyed, as also is taste; respiration through the nose becomes difficult or altogether impossible, and the voice is nasal. The closing up of the nose is usually transient, often sudden, as after blowing the nose; on lying down in bed, the nostril of the side on which the patient rests usually becomes stopped up. If the catarrh extends, as it almost always does, through the nasal duct to the conjunctiva, the latter becomes reddened, and there also occur epiphora, photophobia, and spasm of the lids. When it extends into the naso-pharynx and affects the Eustachian tubes, tinnitus, deafness, and itching in the ear are set up, and perhaps acute otitis media, causing most severe otalgia, which is usually worst during the night.

In infants and very young children the symptoms are more harassing. As they breathe almost only through the nose, their sleep is disturbed, and interrupted by crying and fits of choking; sucking becomes very difficult, as they need to let go the breast as soon as they have taken it, in order to breathe. From want of sleep and insufficient nourishment, the development of the child is interfered with, and if the disease last long, exhaustion and death may occur.

After the symptoms and the secretion have reached their highest pitch, the latter diminishes, as a rule, after from two to four days, becomes thicker, tougher, and extremely rich in mucus, and forms crusts and scabs, or adherent gelatine-like lumps, of a muco-purulent, more rarely purely purulent, nature. The secretion now gradually changes both in quality and quantity till it returns to normal again. The closing up of the nose also disappears gradually, taste and smell return, but not at all seldom the acute catarrh turns into the chronic form.

The objective changes consist in more or less intense reddening of the mucous membrane of both, rarely of one, nostril. The swelling is generally diffuse and uniformly velvety, it is specially marked on the anterior and posterior ends of the inferior spongy bone. There are often erosions and widely dilated bleeding veins. When inflammations are excited by chemical substances, such as bichromate of potash, arsenic, and corrosive sublimate, there very often occur, according to Bécourt, Hillairet, Guay, and Casabianca, deep ulcers of the septum, with perforation of the same, and also ulcerations of the turbinated bones. Perforation of the septum, which, according to Casabianca, often quickly follows, is generally very small, situated $1\frac{1}{2}$ to 2 centimetres above the floor of the nose, and is never followed by sinking in of the nose.

The skin of the nose is reddened, swelled, glistening, and erysipelatous, the skin of the upper lip and nostril is eroded, and often covered with herpetic vesicles. When the nose is completely obstructed, the tongue is dry and fissured, and in consequence of extravasation of blood from the capillaries, has a brownish or black coating. The secretion at the beginning contains desquamated cylindrical and ciliated epithelium, numerous white blood-corpuscles, and mucous cells; the number of the latter increases as the disease goes on, as also does the mucus; very often admixtures of red blood-corpuscles and different kinds of fungi are found.

The course of acute rhinitis extends, as a rule, from three to eight days, but may last for two or even four to six weeks before complete cure takes place. Acute inflammation of the accessory cavities, especially of the frontal sinus, is one of the commonest complications, the symptoms of which will be referred to in the last chapter. More rarely empyema of these cavities, or of the orbits occurs, as was observed by Schäfer and A. Hartmann.

The diagnosis presents no difficulties, once the affection has developed, except in regard to the causation. In the prodromal stage the general symptoms are sometimes so violent that the appearance of an acute disease, an acute exanthema, or of a cerebral affection, is feared.

The prognosis is favourable, though deafness and otorrhœa may result, or the affection may become chronic, and give rise to polypoidal vegetations and neoplasms. A fatal result occurs only in very young children or debilitated old persons, or may be caused by meningitis, as occurred in a case of orbital abscess, recorded by Schäfer.

Treatment.—Treatment is partly prophylactic. As regards removal of the disposition to catarrh, reference must be made to what has been already said about acute pharyngitis. Workmen who have to do with dusty substances, especially the above mentioned chemicals, must protect their nasal mucous membrane with tampons of wadding. Syphilitic or scrofulous patients must at once stop taking the medicine when the first symptoms of iodine catarrh appear. Remedies for aborting catarrh are numberless and much thought of, especially sweating cures, vapour baths and the olfactorium of Hager and Brand; it consists of—

R. Acid. carbol. . . . . .
Liq. ammon. fort. aa, . . . ʒss
Spir. vini. . . . . . . ʒiss
Aq. dest. . . . . . . ʒi

Some five to ten drops are poured on to a small funnel of blotting paper, and inhaled for a few minutes every two hours. Inhalation of chloroform (Solis Cohen), and, according to Unna, the use of a spray of ichthyol. 1, ether., spir. vini. aa, 100 parts, for fifteen minutes, decidedly cut short the course of the catarrh. Lower and Williams recommend immediate abstinence from all fluids, Bresgen the use of apomorphia, gr. $\frac{1}{60}$ to $\frac{1}{12}$ three or four times a day, as a pill. Sulphate of atropia, gr. $\frac{1}{40}$ two or three times a day, as a pill, is also said to shorten and mitigate the affection. Locally, weak solutions of permanganate of potash, and inhalations of oil of turpentine, or 45 to 60 grains of triturated camphor (Dobson) in hot water, insufflations of salicylic acid, or of morphia (gr. $\frac{1}{6}$ for a dose), and, according to Michel, nitrate of silver (1 to 20 of talc.) may be applied to the nasal fossæ. Nasal douches are generally badly borne, and make the trouble worse. When the accessory cavities are also affected, ice to the forehead or nose, leeches to the nostrils or dorsum of the nose, rest in bed, suitable diet, attention to the bowels, and if necessary, opening of abscesses are indicated; in severe conjunctivitis a shade must be worn, and sulphate of zinc solution (gr. $\frac{1}{4}$ to ʒi) instilled; in ear affections, the air douche must be applied, and if necessary, paracentesis of the membrane. Infants must be fed with a spoon, or with the stomach tube, and the above mentioned remedies may be required in proportional doses.

It seems to me that this is the proper place to describe the affection known as *hay catarrh*, or *hay fever*.

By hay fever, or early summer catarrh, is understood an affection which manifests itself as acute catarrh of the nose and of the conjunctiva, sometimes with, and sometimes without, asthma, and which owes its origin to the irritation caused by inhaling the pollen of the blossoms of certain plants and perfumes. It is, according to Blackley, principally caused by the pollen of flowering grasses, of rye, oats, and barley; sometimes by the dust of rose blossoms, hence called "*rose fever;*" in America, by that of the ambrosia ambrosiæfolia. Symptoms similar to, or exactly like, those of hay fever, are often observed in apothecaries and druggists, who have to do with ipecacuanha and lycopodium powder, digitalis, hellebore, and other dusty substances. The smell, too, of horses, dogs, cats, guinea pigs, and rabbits causes the nasal mucous membrane of many persons to become irritated. That hay fever is really due to the above mentioned pollen, is proved by its occurrence almost exclusively at the time of the flowering of the plants: in Europe, from the end of May to the second half of July; in America, in August and September. Blackley also found that the intensity of the disease is in direct proportion to the quantity of pollen floating about in the air. Disposition or idiosyncrasy, too, as one may call it, plays the principal part in hay fever; and just as there are persons who cannot eat strawberries, raspberries, or crabs, without getting urticaria, so there are those who cannot walk about in the open air with impunity at the time of grass flowering. The inhabitants of towns are specially liable to be attacked, as is the Anglo-American race, particularly the male sex. According to more recent observation, hay fever seems to occur almost only in persons whose nasal mucous membrane is already in some way altered, *e.g.*, chronic hyperplastic rhinitis, and is therefore to be regarded only as an acute symptom appearing in the course of a latent chronic nasal affection.

The affection generally begins with violent itching sensations in the angles of the eyes, and pains in the nose and naso-pharynx, followed by severe fits of sneezing, with profuse secretion and lachrymation. The stopping up of the nose is often complete, the irritation of the conjunctiva goes on sometimes to chemosis; sometimes there is also fever. The attacks usually last several hours, more rarely days or weeks, stopping altogether when the flowering of grass is past, only to return next year. In the most severe cases, asthma often sets in, having quite the character of bronchial asthma, and may also last for hours, days, or weeks.

The diagnosis depends on the coincidence of the disease with the flowering of grasses, or with the presence of the above mentioned substances and smells, as well as on the symptoms.

The prognosis is so far unfavourable, in that those disposed to the disease are attacked every year; and favourable in so far, that the disease may be definitely cured by judicious treatment of the mucous membrane, which in most cases has long been affected.

Treatment.—The treatment must be prophylactic, medical, and surgical. As preventives, to be recommended are sea voyages, residence at the sea-side or in a town, keeping one's room at the time when grasses flower, wearing a veil or pair of spectacles with broad closely fitting brims, and tamponading the nose. Efficient medicines are those which act on the nerves, and tonics, such as quinine, arsenic, and iron, and specially good are the pills recommended by M. Mackenzie containing zinc. valer. (gr. $\frac{3}{4}$), and assafoetid. (gr. $1\frac{1}{2}$). For catarrh, insufflations of bismuth (gr. $\frac{3}{4}$) and morphia (gr. $\frac{1}{16}$); for asthma, tinct. opii and saltpetre paper are recommended. For chronic hyperplastic changes of the mucous membrane, the remedies mentioned in the next chapter must be made use of.

A modification of acute and sub-acute rhinitis is *purulent nasal catarrh*.

By this is understood not actual blennorrhœa, but rhinitis accompanied from the beginning by purulent secretion.

Specific infectious material which directly influences the nasal mucous membrane, must be regarded as the cause of this affection. Infection with gonorrhœal poison in adults is not so rare an occurrence, the poison being carried to the nose either by the fingers or by the handkerchief, or by unnatural actions, and producing in it the same changes as it does in the urethra or on the conjunctiva. More frequently purulent nasal catarrh occurs in new-born infants, caused by infection from the secretion of the vagina. Morell Mackenzie considers it not impossible that the irritating influence of the atmosphere, as well as the entrance of soapy water into the nose during careless washing, may excite the affection. During the course of acute infectious diseases, especially scarlet fever and diphtheria, also in consequence of the extension of purulent conjunctival catarrh, and finally from unknown influences, cases of purulent nasal catarrh may arise.

The symptoms are the same as of ordinary catarrh, except that the secretion is of a purulent nature from the beginning; sometimes the secretion is fœtid, thin, slimy, or bloody: the eyes and ears often become infected, and the nostrils and upper lip excoriated.

The course of the disease is generally very tedious; the prognosis is only serious in young children.

Treatment.—Successful treatment is scarcely possible without the nasal douche. Chlorate of potash, and solvents, are specially to be recommended; in adults disinfectants, followed by insufflations of boracic acid, nitrate of silver, or alum (grs. 5 to ʒi of starch), and sulphate of zinc (grs. 1½ to ʒi). As the nasal douche is always dangerous to infants and very young children from their inspiring the fluid, M. Mackenzie recommends while cleansing the nose, temporary plugging of the nasal fossæ by means of a small piece of sponge.

CHRONIC RHINITIS.

Synonym.—Rhinitis Chronica.

Previous attacks of acute catarrh, and inflammatory processes which remain after acute infectious diseases, must be looked upon as the commonest of the exciting causes of chronic nasal catarrh. Scrofulous persons supply a large contingent, as not only are they specially predisposed to nasal catarrh, but also because in them acute catarrhs more easily become chronic than in healthy persons. Experience teaches, that persons who are exposed to inhaling organic and inorganic dust, smokers and snuffers, also frequently suffer from chronic catarrh, and that it is specially liable to attack noses that are malformed, or that are narrowed from deviation of the septum. Bresgen emphasises heredity, and also the origin of chronic rhinitis in consequence of adenoid vegetations and tonsillar hypertrophy, and many consider that mental and sexual excitement, menstruation, and pregnancy are answerable for the disease.

The anatomical substratum of chronic rhinitis is hypertrophy of the mucous membrane, which may be affected either as a whole, or only on individual parts; the sub-mucous connective tissue and glands being mainly affected. Exceedingly often it goes on to tumour-like formations and excrescences, which are very like real polypi *(Rhinitis chronica hypertrophica).* The hypertrophy of the mucous membrane may last for an unlimited time, and may also end in atrophy *(Rhinitis chronica atrophicans).*

The slightest degree of *Rhinitis chronica hypertrophica* manifests itself as reddening and velvet-like swelling of the mucous membrane. It attacks the septum and the turbinated bones, generally both; nowhere is there more pronounced thickening of the mucous membrane present, since generally the objective condition fully coincides with that of acute rhinitis.

When the affection is of very long duration, however, even more pronounced hypertrophy of the mucous membrane takes place.

The inferior turbinated bone is by far the most frequently attacked, its anterior end is changed into a bluish-red, hemispherical or pyriform, smooth tumour, which obliterates more or less the inferior meatus, impinges upon the septum, and may even protrude from the nostril. The posterior end is altered in a similar manner; it resembles a pale gray or grayish-red bladder-like swelling, the base of which occupies the whole extent of the inferior turbinated bone. This tumour occludes the nasal fossæ and extends far into the nasopharynx; it is often very ragged or warty, or resembles a raspberry or piece of frog spawn. Along with the posterior end, often however alone, the middle part of the bone, especially the free border looking downwards and inwards, becomes affected; its mucous membrane then, either hangs down on to the floor of the nose like a smooth bluish-red or pale sack, or appears indented, furrowed, and distinctly papillomatous. The hypertrophied portions are very often distinctly raised up from their surrounding parts. The middle turbinated bone is nearly as often attacked as the inferior; its mucous membrane, especially that of the anterior end, is sometimes uniformly thickened, smooth and globular, sometimes uneven, rough, granular, warty, and polypoid. The mucous membrane of the septum, too, is often the seat of circumscribed or diffuse hypertrophy, sometimes at the part where it joins the floor of the nose, sometimes in the middle, opposite the inferior end of the middle spongy bone, and sometimes at the posterior part of its inferior extremity. The floor of the nose is least often changed, occasionally its hypertrophied mucous membrane touches the inferior turbinated bone, but it very seldom becomes warty or granular.

It appears that in some cases, the nasal bones and cartilages participate in the hypertrophy. According to Brison Delavan, hypertrophy of the spongy bones proper is quite a common occurrence, especially when the septum is malformed. I have several times seen hypertrophy of the cartilaginous septum develop under my eyes; according to Gottstein, the septum is often the seat of chronic perichondritis, which runs an exceedingly slow course, and leads to hypertrophy.

The secretion exhibits manifold variations. In recent cases, accompanied by velvety swelling of the mucous membrane, it is thin and watery, and often as copious as in acute coryza. In other cases it is less plentiful, tenacious, viscid, and extremely rich in mucus, varies in colour from yellow to green and is more muco-purulent; it dries up on the nostrils, septum, and anterior end of the spongy bones, and forms yellow crusts and scabs, which, when forcibly removed, cause bleeding, and lead to excoriations and inflammatory infiltration of the alæ nasi. Pure pus or a fœtid secretion, as far as I have seen, occur but seldom in the idiopathic hypertrophic form.

A frequent accompaniment of chronic rhinitis is chronic catarrh of the naso-pharynx, with or without hypertrophy of the pharyngeal tonsil; other complications are pharyngitis chronica, hypertrophy of the tonsils and lateral folds of the pharynx; the cavities adjacent to the nose, particularly the maxillary and frontal sinuses, sometimes participate in the inflammation.

The symptoms of chronic hyperplastic rhinitis principally show themselves in hindrance of nasal respiration. The obstruction of the nose is lasting or more often temporary, and is caused partly by the accumulation of secretion, partly by the swelling of the erectile tissue of the inferior spongy bone. The more the nose is obstructed, the greater is the disturbance of smell and taste; nevertheless even in partial swelling, especially of the olfactory region, complete anosmia (loss of smell) may take place. Nasal timbre of the voice, holding open the mouth, dryness of the tongue, with mawkish taste, and restless sleep disturbed by wild dreams, are further consequences of nasal obstruction. Very often there are disturbances of hearing, varying in degree from tinnitus to severe inflammation of the middle ear. When the pharynx also is affected, choking and vomiting, chiefly in the morning, and sometimes cough occur. Removing the copiously secreted viscid mucus gives very great trouble, and blowing the nose becomes painful when the alæ are excoriated and infiltrated. Repeated, and sometimes very profuse, epistaxis occurs, either in consequence of violently blowing the nose, snuffling and rubbing, or spontaneously, in consequence of congestion, a symptom which has rightly been pointed out by Bresgen as being very important, and which is often overlooked. Swelling and reddening of the point and dorsum of the nose, to which we shall presently allude, are also frequent accompaniments of chronic rhinitis. It is generally a result of engorgement, occasioned by pressure on the turgid spongy tissue, which prevents the normal return of the blood.

Among other symptoms which are almost constantly present, are heaviness of the eyes, giddiness, aversion and inability for mental work, failure of the memory, sneezing, irritation, and itchiness in the nose.

Reflex neuroses belong to the rarer sequelæ of chronic nasal catarrh. By this term are understood symptoms which occur quite as much in the diseased nose as they do in other organs more or less remote from it. Voltolini, many years ago, called attention to the relations between asthma and nasal polypi, and quite lately Hack has proved the connection of many other neuroses with diseases of the nose. His observations meanwhile have been to a large extent confirmed by Bresgen, Sommerbrodt, Schäffer, Eugen Fränkel, the author, and others.

If we turn first to the local reflex symptoms affecting the nose, we meet with the so-called "*nervous catarrh*," in which affection an extremely profuse, clear, thin, or tough secretion makes its appearance suddenly and at intervals. Diminution or complete loss of smell, obstruction of the nose, and violent sneezing are usually associated with it. The attacks often last several hours, and are most violent at the time of menstruation. Besides the secretion from the nose, lachrymation usually occurs, and is sometimes combined with scintillating scotoma. Hay fever, too, has generally its ultimate cause in chronic rhinitis. A very rare reflex neurosis is ptyalism, observed by E. Fränkel and myself. Among reflex neuroses producing results nearest the original seat of disease, are neuralgiæ of one or several branches of the trigeminus, especially of the supra-orbital. The different kinds of headache, megrim, occipital, and forehead pain, and dull pressure on the vertex, also belong to this category. The erythematous reddening and œdematous erysipelatous swelling of the external nose, cheeks, eyelids, and conjunctiva, which often suddenly appear and as quickly disappear, are to be considered as a vasomotor neurosis. They set in sometimes on the slightest irritation affecting the nose, or on taking very hot foods or drinks, and in a few hours disappear, leaving no trace. Their frequent return may lead to permanent changes of the cutis, the well-known " Kupfernase " (gutta rosacea).

It is also probable that vertigo and epileptic attacks may be excited by nasal affections (Hack, Löwe); and there is no doubt that cough is. It begins with excessive tickling irritation in the larynx, generally at night, sometimes lasting only several hours, sometimes going on continually, and being characterised by the absence of expectoration and perfect soundness of the lungs and bronchi. E. Fränkel considers it further probable, that a large proportion of cases, the etiology of which is very hazy, and which are set down as neuroses of sensibility, might, on stricter examination, be traced to diseases of the nose. Sometimes the irritation proceeding from the nose attacks the cervico brachial plexus, and excites neuralgic pain in the upper arms, darting pains in the breast, and between the shoulder blades. Sommerbrodt observed vasodilator reflex phenomena affecting the mucous membrane of the bronchi,—bronchitis with sibilant rales without asthma,— also reflex phenomena which caused vomiting, and phenomena affecting the skin,—chilliness and shivering with pallor.

One of the most frequent and most painful reflex neuroses is *asthma*. Formerly it was believed only to occur when neoplasms were present, now we know that it appears just as often in chronic inflammation of the turbinated bones. As a rule the attacks come on at night, while or after falling asleep, last for several hours, and towards morning gradually pass off. Some patients suffer from these attacks every night, others only every other week, and are perfectly well during the day. Others again experience a certain feeling of oppression during the day. For some time or immediately before the onset of asthma, there is very often a profuse watery secretion from the nose and sneezing. I myself am certain that bronchial spasm is the cause of asthma, and that Curschmann's *bronchiolitis exudativa* is the result and not the cause of asthma. Disease of the nasal mucous membrane may set up irritation of the vagus as directly as it may arise from overloading the blood with carbonic acid, in consequence of the erectile tissue covering the turbinated bones becoming swollen during sleep.

As we must again refer to asthma in connection with neoplasms of the nose, let me here mention in conclusion, that I once observed a case in which the patient experienced sensations of forced movement in consequence of chronic naso-pharyngeal catarrh.

Among the exciting causes of reflex neuroses, the principal are, influences of temperature, exposure to bad air, close and overheated rooms, or rooms filled with dusty substances, smells of all kinds, good and bad, the odours of certain perfumes and plants, especially of hay or flowering grasses, roses and carnations, and also psychical disturbances and menstruation. Not only are more or less pronounced changes of the mucous membrane always necessary for the production of reflex phenomena, but also an unusually heightened excitability of the nervous system. According to E. Fränkel these sometimes appear congenitally, or develop during the first year of life. More often they develop as results of measles and whooping-cough, or in consequence of adenoid vegetations. According to Hack, the spongy tissue, especially that of the inferior turbinated bone, is observed as the place from which the reflex phenomena arise; they may be more easily excited in insignificant and strictly localised processes, than in diffuse hyperplastic rhinitis; if it were possible to remove the spongy tissue, the symptoms caused by it would disappear. Although now I believe that the spongy tissue plays no insignificant part, especially in the origin of nocturnal asthma, still it appears to me, that too much significance is ascribed to it. The production of reflex phenomena certainly takes place from the sensory nerves of the mucous membrane, usually without the cavernous reticulum being affected, an opinion which was generally accepted at the Copenhagen Congress.

The diagnosis of rhinitis hypertrophica presents no difficulties when a careful examination is made; the use of the probe is absolutely necessary for determining the diagnosis; each part of the mucous membrane that looks doughy and soft, that can be more deeply probed, or be pushed backwards and forwards, or otherwise out of its place, must be regarded as morbid, that is to say, hypertrophic. Bresgen is perfectly right when he says, that up till now far too little importance has been given to this circumstance, and that very often chronic rhinitis has been overlooked. It is easy to understand that the diagnosis becomes all the easier, the more pronounced the changes appear, especially on the ends of the turbinated bones. The condition of the spongy tissue is of diagnostic value in this respect, that its increasing and diminishing on slight irritation must be regarded as a symptom of increased reflex excitability, just as much as the occurrence of attacks of sneezing, lachrymation, or coughing, when the rest of the mucous membrane is stimulated.

On the appearance of the above-mentioned neuroses, a most careful examination of the nose should never be omitted, even although no symptoms pertaining to it are complained of. Now that the connection of so many different neuroses with nasal affections has been irrefutably demonstrated, it is the duty of the conscientious physician, either himself to examine with the rhinoscope every patient affected with asthma or emphysema, or get them examined by a competent person. The presence of diseases in other organs should never be allowed to overrule the inspection of the nose, even when they can fully account for the symptoms. I am here forcibly reminded of a case recorded by Eugen Fränkel, of a lady with asthma, who was suffering from left-sided valvular disease, and who was entirely cured of her asthma by treatment of a nasal affection, which at first was looked upon as of quite secondary consideration.

But I should also like earnestly to warn the beginner against hasty diagnoses and speculations. Apart from the fact that a long time must elapse before the doctrine of reflex neuroses can be regarded as settled at present we do not possess a single criterion, which enables us to determine with certainty, whether the neurosis under consideration is really connected with the existing disease of the nose.

The prognosis of chronic hyperplastic rhinitis is generally good, as with sufficient perseverance the patients can, to a large extent, be cured of their symptoms, particularly those caused by obstruction. With regard to the removal of reflex neuroses, the prognosis is doubtful; I can only strongly advise, that it be briefly explained to the patients, who generally have had much treatment and are filled with exaggerated hopes, that it is considered that treatment of the diseased nose is indicated, but that the results cannot in any way be guaranteed. There is no doubt, and I have already mentioned, that the majority of reflex neuroses dependent on nasal disease admit of a favourable prognosis, nevertheless it must not be forgotten, that bad results are not so rare, as the large number of hitherto recorded favourable cases might perhaps lead one to conclude. The relatively best prognosis in my experience may be given in affections of the trigeminus, the different forms of headache, and hay fever, less favourable in red nose, and relatively most unfavourable in asthma and disturbance of smell. With regard to asthma, I must certainly mention that in most of the patients treated by me and others, especially by Schäffer and Eugen Fränkel, the trouble had existed for many years before the nose was treated; therefore it is to be hoped that the prospects of radical cure of so distressing a disease will become more favourable, the earlier in future the patient seeks the advice of the physician. Cure seldom occurs suddenly or all at once, as a rule it requires repeated interference, and very long and careful local, and particularly general, treatment. Asthmatic persons often improve so far, that the attacks occur more seldom and are less severe, or else assume a different character.

Treatment.—For milder forms of chronic rhinitis, insufflations of silver nitrate are recommended, after cleansing the nose with some solvent. On account of the great sensitiveness of the nose. much caution, as Bresgen rightly affirms, must be used as to the dose of this remedy. At first a mixture of silver nitrate, grs. 2½, to starch ʒi, is enough, gradually increased to grs. 5–25 to starch ʒi. A portion of this powder, generally grs. 10, is insufflated into each nostril alternately, as the reaction would be too severe if both were treated at once. Other astringents, such as alum, bismuth, tannin, zinc acetate, and zinc sulphate, are less efficient. Porter has had very favourable results from the insufflation of a powder consisting of camphor, salicylic acid, and tannin; Gottstein from that of salicylic acid, 1 to 10 of magnesia usta; Morell Mackenzie recommends the nasal douche with the "compound alkaline wash," which is prescribed in the form of a powder, thus :—

  ℞ Sodae bicarb.
   Sodae biborat.
   Sodii chlorid. aa, ... grs. vii.
   Sacch. alb. . . . . . . grs. xv.
  Sig. To be dissolved in half a tumblerful of tepid water.

Nasal douches with tannic acid, grs. 1½ to ʒi of water, or with alum, grs. iii. to ʒi of water, and the tannin spray, grs. iii. to ʒi of water, are considered by Mackenzie to be specially effective. Medicated bougies are not only more troublesome, but less effectual. Dried up crusts and scabs are softened with vaseline or almond oil, excoriations are touched with solid caustic, or painted over with a solution of iodine in glycerine; for excoriations and infiltrations of the nostrils, I can best recommend the application of little plugs of wadding impregnated with boracic acid in vaseline.

Although very effectual in the milder form, these remedies are of little or no use in severe cases with considerable hypertrophy. In such cases no time should be lost in applying the only effective remedy, the galvanic cautery.

Polypoid vegetations, which are to some extent pedunculate, may be removed either with the cold or the electric snare. The removal of flat and extensive vegetations, and diffuse hypertrophies of the mucous membrane, is generally very difficult and tedious. While the simple passing of a flat cautery over the slightly swelled ends of the spongy bones is usually sufficient, the same treatment in well marked cases appears to be totally insufficient; on hypertrophy of the posterior ends of the spongy bones, I operate with the electric snare only, and similarly on the portion of mucous membrane hanging down on to the floor of the nose. Although the amputation of the spongy tissue sometimes causes considerable hæmorrhage, hitherto I have always been able to control it in a short time. Hack recommended puncture and furrowing with the galvanic cautery in order to avoid hæmorrhage. The former is done with a pointed cautery, which is stuck deeply and repeatedly into the parts to be destroyed, while the latter method requires it to be entered at one place only, and from there drawn through the whole diseased part, while at a glowing heat.

The amputation of the posterior ends of the spongy bones is a very difficult manipulation. In America, this is accomplished with Jarvis's nasal ecraseur, while in Germany the cold snare is used. I myself recommend in all cases, on account of its safety, that the electric snare be passed through the nose over the parts to be removed, the wire allowed to get slightly heated, and then tightened by pulling back the slide. In this way the wire tends to adhere, and is prevented from slipping off the thickened and slippery mucous membrane covering the extremity of the turbinated bone. If it cannot be seized by the snare, a suitably bent cautery with a broad point must be energetically and repeatedly bored into the tissue. The galvanic cautery, where it is required, must not be used timidly, the simple and too often single application of it to the diseased mucous membrane, is now condemned by Hack as insufficient and even hurtful, since only the complete destruction of the hypertrophied parts can guarantee permanent and total cure, especially of reflex neuroses. The protective apparatus recommended in several quarters, nose shields, &c., not only obstruct the field of operation, but are also completely useless for those in any way skilled in operating.

With regard to bad complications, reaction, and after treatment, I must refer to the chapter on neoplasms. Of course, pharyngeal catarrh, adenoid vegetations, scrofula, and anæmia must be treated on sound general principles. In obstinate cases of red nose, I have had excellent results from puncturing the skin of the nose all over. If difficulties in hearing do not soon disappear, the air douche must be applied to the middle ear. More efficient against asthma than all remedies acting on the nerves, and specific palliatives, such as quinine, bromide and iodide of potassium, Fowler's solution of arsenic, Aubry's arcanum, saltpetre paper, chloroform, pneumatic apparatus, morphia, atropin, and opium, is the application of Faradism to the vagus externally in the neck, below and within the angle of the jaw, and each application must last for at least a quarter of an hour, better still for from half an hour to one hour.

We must now here consider *Rhinoscleroma*, a chronic inflammation also depending on hypertrophy of the mucous membrane.

While by Hebra and Kaposi it is discussed among the neoplasms, and by others among the results of syphilis, it is looked upon by Billroth, Geber, and Mikulicz, as an extremely tedious inflammatory process, which, after small celled infiltration of the superficial and deep layers of the skin and mucous membrane, leads to wrinkling or shrivelling of the connective tissue.

Its etiology is altogether obscure, in several cases there were injuries to the external nose before the disease broke out. Constitutional diseases, such as syphilis, scrofula, and tuberculosis, do not appear to influence its origin, and it is also very doubtful whether the bacteria discovered by Frisch in rhinoscleroma are really the cause of it.

Rhinoscleroma almost always begins on one of the alæ nasi, or on the nasal septum; sometimes, however, in the pharynx, on the soft palate and in the naso-pharynx. As the disease advances it spreads to the lips, gums, and cheeks, and generally the whole extent of the internal nose is attacked; after the pharynx is affected, the larynx also in the end becomes involved. The disease appears as flat or elevated, sharply defined or diffuse nodes and patches, which are extremely hard to the touch and painful on pressure. Their colour is either the same as that of the surrounding skin or mucous membrane, or varies from clear to dark bluish-red, and glistens like a hypertrophic cicatrix. It is characteristic that the nodes do not necrose; at most they only become excoriated superficially. After being excised they quickly reappear. The soft palate becomes changed into a glistening band of cicatricial tissue, while the epiglottis appears as a blunted cone.

The subjective symptoms consist, besides disfigurement of the face, in displacement of the nose, pain on pressure, and narrowing of the oral aperture; when the pharynx is affected, in anginous speech, and difficulty in swallowing; and when the larynx participates, in hoarseness and dyspnœa, the general health not being usually disturbed.

The diagnosis is somewhat difficult, as rhinoscleroma bears a very striking resemblance to syphilis, especially when the nodes become superficially excoriated, and when the pharynx also is affected; the very extraordinary hardness of the nodes, and their not becoming soft, as well as their not being influenced by anti-syphilitic treatment, must strengthen the suspicion of rhinoscleroma.

The prognosis is unfavourable, as the affection always returns, and may directly threaten life by involving the larynx.

Treatment—Treatment is rather hopeless, since general treatment is of no use, while local remedies may produce improvement for a time, but never cure. The obstruction of the nose requires dilatation by means of catgut, bougies, or laminaria, and caustic potash, chromic acid, and the galvano-caustic point may be used to destroy the tubercles.

Chronic inflammation of the mucous membrane of the nose, like that of the pharynx, may terminate in atrophy (*Rhinitis chronica atrophicans*).

Whether atrophy is always the last stage of hyperplasia, or whether it may not, from the very beginning, attack the healthy mucous membrane, is not yet decided, nevertheless numerous observations go to prove that, in the majority of cases, it is preceded by an hypertrophic stage. Atrophy attacks all the constituent elements of the nasal mucous membrane, epithelium, spongy tissue, glands, and even the bones. The glands diminish in number and size, and the mucous membrane becomes altered into fibrous connective tissue.

The most constant and most certain objective symptom of atrophic rhinitis is the abnormal roominess of the nostrils. The spongy bones, especially the inferior, appear unusually small, and changed into small round or flattened crests, and the nasal meatus unusually roomy; through atrophy of the cavernous tissue the view into the naso-pharynx is rendered extremely easy, and, as a rule, one sees not only the Eustachian prominences, but also the openings of the tubes and a large portion of the posterior wall lying between them.

The atrophied mucous membrane, cleared of secretion, is sometimes slightly reddened, often pale, or of a normal colour. Ulcerations or destruction of the cartilages and bones do not occur; at most there are small superficial ulcers or excoriations of the nasal apertures, or of the septum, as in all forms of rhinitis.

The constant accompaniment of atrophic nasal catarrh is pharyngitis sicca, which, although generally localised to the naso-pharynx, may also extend to the pharyngo-oral cavity.

A peculiar modification of chronic atrophic rhinitis is *blennorrhœa of the respiratory mucous membrane*, described by Stœrk. This affection occurs endemically in Bessarabia, Gallicia, Poland, and South

Russia. It begins gradually in the nose, and often, after many years, spreads to the pharynx, larynx, trachea, and even to the bronchi, where, by proliferation and retraction of the connective tissue, ossification of the cartilages and adhesion of the vocal cords, it may cause stenoses dangerous to life.

The secretion in atrophic rhinitis is always purulent, and when recent is yellow, green, or cream-like, hence the disease is also called nasal blennorrhœa. It possesses in a marked degree the power of rapidly drying up into hard yellowish or dirty green, or brown crusts, as long as a finger, which cover the whole of the nasal mucous membrane like a carpet, and emit a very characteristic penetrating fœtid odour. Although every atrophic nasal catarrh is not accompanied by fœtor, still, in the majority of more fully developed cases, it is never absent. This is the form which previously was called *ozæna*, and now is known as "*genuine ozæna*," or *rhinitis chronica atrophicans fœtida*. Latterly there has been much lively discussion over the term "ozæna." Although it appears unsuitable to name a disease, on account of one prominent symptom, which is also observed under other conditions, especially in processes destructive to mucous membrane, cartilages, and bones, yet, in my opinion, there is no reason why the term ozæna should be entirely abolished, for if it be limited to rhinitis chronica atrophicans fœtida, it can give rise to no further misunderstanding. The secretion of genuine ozæna is considered by some to be infectious, but as up till now I have never observed a case of infection, I must deny that it is so. This power of infection is proved only in the case of Stœrk's blennorrhœa, which owes its origin and spread to the most unfavourable hygienic and social conditions, and to the incredible dirtiness of the population of the above mentioned countries.

In the etiology of genuine ozæna, anomalous states of the blood and dyscrasic conditions hold the first, if not the only place. This is especially true of the scrofulous, a class whose compass must certainly be held to extend further than to those individuals who are the subjects of swollen nose and upper lip, skin eruptions, and enormously enlarged lymphatic glands. When it is declared, that perfectly healthy persons also are attacked by genuine ozæna, it is only partly correct, because on closer examination it is found that the diathesis of these persons leaves much to be desired. I at least do not remember a single case of genuine ozæna, in which the patient did not exhibit traces of anæmia, chlorosis, scrofula, or tuberculosis.

Genuine ozæna often develops after measles and diphtheria, or after typhoid and severe parturition, therefore after those diseases which are associated with anomalous conditions of the blood and of nutrition. I must decidedly deny that expanded nostrils, their being directed forwards, the absence of hair in the nostrils, curvatures, outgrowths, and perforations of the septum, are the causes of genuine ozæna. The female sex is particulary predisposed between the ages of six and seventeen, that is, during the time of puberty; only in two cases have I seen the affection develop after the twentieth year: once in a highly chlorotic lady, after the removal of numerous polypi which obstructed the nose; the other case was that of an elderly married woman, who, in consequence of a uterine disorder, was very much pulled down.

I have repeatedly seen atrophic rhinitis, with or without characteristic fœtor, follow cure of nasal syphilis, and I am inclined to put it down to retraction of the gummatous infiltration and cicatricial changes in the connective tissue of the ulcerated parts.

No question has of late years been oftener or more warmly discussed, than that of the cause of the fœtor.

Some authorities consider abnormal width of the nostrils, caused by atrophy, to be the cause of the smell, and as a proof, point to the rapid drying up and difficult removal of the secretion, which by remaining there becomes fœtid; that the width of the nostrils alone cannot be the cause of fœtor is proved by the fact, that very often in spite of it, no fœtor is present. Others look upon atrophy of the mucous membrane and fatty degeneration of the mucous corpuscles with formation of fatty acids as the cause of the fœtor; to this is opposed the fact, that the fœtor may be observed not only when the mucous membrane is atrophic, but also when hypertrophic. According to Ziem, if a ferment be necessary for the origin of the smell, then the overloading of the nose and its accessory cavities with putrefactive gases, may be considered the cause of the fœtor. Others again, think that the micro-organisms which develop in the dried secretion are the causes of decomposition and of the fœtor.

I myself believe, that in the physical and chemical changes of the secretion caused by atrophy of the mucous membrane, is given the condition necessary for the occurrence of fœtor. I have never as yet at least seen genuine ozæna, in which atrophy of the mucous membrane and abnormal width of the nose were not present. I am, therefore, very suspicious of "genuine ozæna," when it is said to be often associated with hypertrophy of the mucous membrane, and without wishing to doubt that atrophic patches, near or beside the hypertrophic portions of the mucous membrane, may be present from which the fœtor arises, I must, however, lay down this advice, founded on many years' observation directed exactly to this point, that such cases are much more likely to be due to syphilis, a foreign body, concretions, or parasites, particularly, as we shall see later on, as syphilis so often appears in very disguised forms.

The subjective symptoms of atrophic rhinitis consist in disturbances of smell, headache, pressure on the eyes, nasal timbre of the voice, difficulty in hearing and tinnitus aurium, while dysphagia and hawking, in consequence of dry naso-pharyngeal catarrh, which is always present, are quite common. With regard to nasal obstruction, numerous differences and many variations are met with; generally the nose becomes impermeable only when the dried up secretions and crusts are very abundant, and becomes free again whenever they are removed. By forcible attempts to remove the secretion by blowing the nose and picking it with the fingers, hæmorrhage is caused, which imparts to the crusts a brownish black colour. But no symptom is so characteristic as the fœtor; to compare it with other smells, as for example, with that of old cheese or of crushed bugs, is quite impossible; it is indeed so peculiar, that whoever smells it once will always recognise it again. While, as a rule, the smell remains quite unperceived by the patients, it becomes all the more noticeable to those around, so that the miserable sufferers are avoided and often expelled from their social position, and may also suffer as regards their means of livelihood. Under such circumstances, the training of young girls,—for example—is scarcely to be thought of. The blennorrhœa of Stoerk, which is always accompanied by fœtor, leads later on in its course to hoarseness, loss of voice, and apnœa.

The diagnosis of atrophic rhinitis is founded on the one hand, on the objective and subjective symptoms mentioned, on the other, on the absence of deep or extensive ulcerations of the mucous membrane, but particularly on the non-participation of the cartilages and bones. If the latter are affected, then we can no longer consider the case to be one of genuine ozæna, but as one of syphilis or some other of the severe affections formerly designated as "*symptomatic ozæna.*"

The prognosis of rhinitis atrophicans, and especially of the fœtid form, is absolutely hopeless, as we possess no remedy which can restore the mucous membrane to its normal state, and radically and permanently remove the fœtor. Contrary assertions are founded on diagnostic errors, although there is no doubt that fœtid catarrhs with hypertrophy of the mucous membrane are cured either spontaneously or by artificial means. One should make it a rule to inform the patient's friends at the outset, that true ozæna is incurable, but should also call their attention to the fact that by artificial means the most troublesome symptom—the fœtor—can be completely concealed and partly removed, and that otherwise the malady has no injurious influence on the duration of life. In this last respect Stoerk's blennorrhœa offers gloomy prospects, especially in cases of long standing, combined with laryngeal and tracheal stenosis.

Treatment.—In the treatment, the first and foremost point to be considered is the removal of the secretion. This is accomplished by means of the nasal douche, and as a larger quantity of water and more pressure than usual are required, on account of the thickness and toughness of the crusts, an English syringe, or a syringe with piston action, is the best to use. For cleansing and disinfecting I use either chlorate of potash (a teaspoonful to a litre of water), or the permanganate (grs. 5-10 to a pint of water), while in non-fœtid rhinitis 1 to 2 per cent. solutions of equal parts of carbonate of soda and chloride of sodium.

Scabs and crusts, especially those on the roof of the nose and on the middle spongy bone, which resist the douche or the spray, must be removed with a pair of forceps, or with Hartmann's nasal brush. Cleansing the nose twice daily—morning and evening—is generally sufficient. The other disinfectants, such as boracic acid, carbolic acid, resorcin (grs. 1½ to ʒi), thymol, benzoate of soda, and salicylate of soda, remove the fœtor to a certain extent, and are quite indispensable when the disease lasts long. In America, douching the nose with a large nasal spray is preferred, especially by Lefferts, and for this purpose "Listerine" is used, a remedy consisting of thymol, eucalyptol, mentha arvenis, gaultheria, &c.

The second point to be aimed at is prevention of the formation of new crusts. Unfortunately we can do so only to a limited degree. It is best accomplished by tamponading, recommended by Gottstein, and the insufflation of stimulating remedies.

When Gottstein's tamponade is called anti-hygienic, fœtor preserving and objectionable, it only goes to show that its purpose is entirely misunderstood. The purpose for which it is intended—viz: to produce a more copious and more fluid secretion by slightly stimulating the mucous membrane—is, however, most completely accomplished. The method is as follows:—the secretions being entirely removed, a piece of Brun's cotton wool, about the length of a finger, and from a half to one centimetre thick, is rolled round a small screw, ending in a stem fixed to a handle, which is introduced into the nostril with a rotatory movement; the screw is now reversed and withdrawn, leaving the plug behind, which, by means of a probe, is pushed as high as possible into the nose, where it is allowed to remain for several hours; if necessary, one side may be treated in this way during the day, the other during the night. Salicylic wool, or wool impregnated with iodoform or with other materials, have no real advantage. Patients soon learn to apply tampons themselves.

Copious secretion can also be produced by insufflating stimulating powders. The best of these is nitrate of silver (grs. 5–50 to ʒi of starch). In England and America, red gum (one part to two of starch), galanga and sanguinaria are much preferred. With us, disinfectants, especially pure boracic acid, salicylic acid, iodoform, resorcin, and in pronounced cases of scrofula, the formula calomel ʒi, hydrarg. præcip. rubr. ʒss, amyl. ʒi, is applied after the removal of the crusts.

Ziem, relying on his theory of the overloading of the nose with decomposition gases, recommended ventilation of the nose as the simplest treatment. This consists in the patient, after removing any tight clothing, going out every two hours into the fresh air, where possible, by the sea side, or into a wood, and there inspiring and expiring as deeply as possible through the nose. Besides this, he must inhale, with as deep inspirations as he can, a 1 p. c. solution of common salt, or of pine oil, or of eucalyptus oil from the leaves (Ol. eucalyp. min. 5–15, spir. vini. ʒi, aq. dest. ʒi). Stumps of bad teeth must be removed, and any hypertrophic parts of the mucous membrane destroyed with the galvano-caustic point.

Whether or not the application of the galvanic cautery which B. Fränkel recommends, exerts a favourable influence on the secretion of the atrophic mucous membrane, I do not venture to decide, although several observations by M. Mackenzie point to its doing so.

The third indication in the treatment is the improvement of the constitution, by means of cod liver oil, quinine, iron, syrup of the iodide of iron, nourishing diet, and residence in the country, or at the sea side.

In the case of Stoerk's blennorrhœa, treatment, besides the above mentioned methods for the nose, consists in cauterising the larynx with nitrate of silver, separating the adherent vocal cords, in excision of cicatricial bands which give rise to stenosis, and in the insertion of Schroetter's bougies. When adhesion again takes place, as it often does after a very short time, nothing remains but to perform tracheotomy, which, however, is of no use, if the deeper parts of the trachea are also stenosed.

PHLEGMONOUS RHINITIS.

Synonym.—Rhinitis Phlegmonosa.

Although the nose, on account of the very intimate attachment of the mucous membrane to the cartilages and bones, is but little predisposed to phlegmonous inflammations, still now and again they have been met with.

The cause depends either on an exacerbation of acute catarrh, infection with putrid matter, or on injuries to the mucous membrane, in consequence of foreign bodies, wounds of the external nose, and operative procedures. It is well known that facial erysipelas often begins in the nose, also that phlegmon tends to associate itself with other diseases and affections of the cartilages and bones, especially when these are ulcerative. Sometimes phlegmonous inflammation of the accessory cavities, particularly empyæma of the maxillary sinus, spreads to the nasal cavities.

The septum appears to be the part which is chiefly attacked. I have often seen at the very commencement of erysipelas, the mucous membrane of the septum intensely reddened, swollen, and covered with vesicles, or on it are formed circumscribed roundish swellings, which block up more or less one or both nasal cavities, and even protrude from the nostril. Such swellings are felt to be soft and œdematous, and show distinct fluctuation when they come to suppurate. The external covering of the nose, and that of the cheeks and upper lip also, are affected by the inflammation, the nose appears swollen and reddened, is very sensitive and painful to the touch. Phlegmon very seldom indeed develops on the spongy bones, but the entrance to the nose is very often attacked by circumscribed phlegmon and follicular abscesses. These almost always originate in the hair follicles, and run a very chronic course.

The subjective symptoms are fever, burning, beating, and darting pains in the nose, pain in the forehead, tinnitus, giddiness, sometimes real meningeal symptoms, vomiting, dulness of consciousness, &c., also stopping up of the nose and increased secretion.

The course is generally rapid. The process may end in resolution within a few days, especially when suitable treatment is applied, or may end in abscess formation. Chronic abscesses also occur. A fatal issue very rarely takes place from meningitis. After the abscess has been opened, the affection may go on to deformity of the nose, caused by caries and necrosis, particularly in those forms which originate in the bones.

Treatment.—Treatment from the outset must be energetic and antiphlogistic. The spread of the inflammation may often be prevented, and the symptoms mitigated by applying bladders and compresses of ice, and by syringing with cold water. Whenever fluctuation appears, the abscess must be opened. If suppuration is protracted, injections of warm water, diluted milk or camomile tea, and moist warm poultices, or inhalations of warm vapours, are indicated. After opening the abscess, douching with disinfectant solutions is recommended.

DIPHTHERITIC RHINITIS.

Synonym.—Rhinitis diphtheritica.

Diphtheria of the nose is almost always secondary, that is to say, communicated from the pharynx. The nose is very rarely attacked primarily. When such is the case, erosions, ulcers, or recent wounds generally exist previously. The case is related by Johnston of a lady, who was seized with diphtheria of the nose and naso-pharynx, on the third day after operation for nasal polypus, after she had visited her dressmaker, who was probably suffering from diphtheria.

The way in which infection takes place is sometimes very obscure. Schuller, among other cases, mentions that of a child five weeks old, who, from birth, was affected with diphtheritic catarrh, having probably been infected during birth by the diphtheritic vagina.

According to Monti, nasal diphtheria is not so very rare a disease among newly-born children and sucklings in the first six or eight weeks of life. It generally appears independently of any diphtheritic epidemic, and its occurrence can but rarely be traced to diphtheritic contagion. Monti considers it probable that, in the majority of cases, the infection is communicated by some puerperal process in the mother.

Primary nasal diphtheria of infants begins with fever, lassitude, and apathy; they sleep much, and seldom take the breast. The mucous membrane of the nose is swollen and reddened, and secretes a thin watery secretion which erodes the skin of the lips and nostrils. The mouth is held open, and moist râles are heard in the nose. After a fresh attack of fever, which generally occurs in from one to three days, the nasal secretion becomes copious, sanious, and mixed with blood and pseudo-membranes. The affection may now heal, or may spread to the lips, mouth, and pharynx, or may cause death by sepsis.

The prognosis in infants is very bad; only when the exudation is very slight has a cure been observed.

Treatment is the same as in secondary nasal diphtheria (see Diseases of the Pharynx, p. 167).

SYPHILIS.

The nose, like all the respiratory organs, is particularly liable to be attacked by syphilis.

The nose is sometimes primarily affected, by poison from a chancre being carried to the nostrils. The first manifestation of secondary syphilis is erythema or syphilitic catarrh. It occurs particularly often in newly-born infants. There are, however, no distinguishing symptoms between specific and non-specific rhinitis. Papular exanthema also attacks the nose. The chief seat is the cartilaginous septum, the floor, and entrance to the nose, more rarely the spongy bones. According to Bäumler plaques occur preferably in infants and children. Their course is almost exactly identical with those of the mouth and pharynx. Examination reveals roundish spots on the considerably reddened and tumid mucous membrane, over which, after a longer or shorter time, the epithelium becomes raised, and superficial ulcers develop. If some trace back the destructions of the septum and turbinated bones exclusively to disintegrated papules, I must maintain an opposite opinion, although I allow that sometimes the plaques seize on the deeper parts and also affect the cartilages and bones.

I have several times seen syphilitic condylomata at the entrance and on the floor of the nose, and also on the border between the skin and mucous membrane.

The later stages of syphilis far oftener give rise to affections of the nose. Gummatous infiltration of the mucous membrane must be considered the almost constant cause of the frightful destructions so often met with.

At first apparently harmless, and often quite unrecognised, the infiltration runs on under the form of chronic rhinitis, accompanied by obstruction of the nose, disturbance of smell, increased watery mucous secretion, and nasal timbre of the voice; examination at this stage revealing nothing but reddening and swelling. After a longer or shorter time, often after several months, the secretion becomes thicker and more purulent, but still not at all fœtid, and the obstruction becomes complete. Inspection of the deeper parts of the nose, which, on account of the swelling of the mucous membrane, is very difficult, or even impossible, may show that the mucosa is still intact. Generally, however, there are superficial ulcers seen, chiefly on the septum and turbinated bones, which more than strengthen the suspicion of syphilis. If the patients come still later under treatment, then, in addition to the already described symptoms, pain in the nose, fœtid discharge mixed with blood, and sinking in of the bridge or point of the nose, make their appearance. Examination now reveals dirty looking ulcers of different sizes on the turbinated bones or septum, with perforation of the latter, and caries and necrosis of the turbinated bones, which either project freely into the nasal cavities, like blackish-gray masses, or are covered with dirty green crusts, and are sometimes surrounded by numerous polypoid excrescences (ozæna syphilitica).

Although, as a rule, the cartilages and bones are laid bare, still, from observations made by myself, Sänger, and Schuster, primary perichondritis, together with gummatous and absorptive periostitis, are not by any means rare. If these processes are localised in the neighbourhood of the nasal bones, and if the disease attacks the septum from the inner lamella of the ethmoid bone, the bridge of the nose may sink in without previous obstruction and fœtid discharge. The pus, as I once saw, may also burrow under the mucous membrane of the septum to the floor of the nasal cavities, obliterate both nares, and give rise to dangerous facial erysipelas. The perforations of the septum, which usually have their seat in the anterior cartilaginous part, are of very different sizes and forms, though, as a rule, roundish or oval; their edges often ulcerate for a long time, and are constantly covered with crusts, even after their healing; frequently several defects are present.

Gummatous infiltration is very commonly localised near the entrance and alæ of the nose. These parts are seen to be more or less thickened, tense, and nodular, the aperture narrowed, later on ulcerated or covered with cracks and fissures, to which yellowish dry scabs adhere. After the ulcers have healed, that is to say, after the cartilages are destroyed, the alæ nasi sink in, and so give rise to narrowing of the nasal apertures and deformity. Ulcerating infiltration of the floor of the nose may lead to caries and perforation of the hard palate.

The diagnosis of syphilis in the first stage, *i e.*, during the stage of erythema and gummatous rhinitis, is absolutely impossible, so long as there are no ulcerations, and while other symptoms of the dyscrasia are absent. The diagnosis of plaques is easy, especially when they are also present in the mouth and pharynx. Obstinate catarrh is, in my experience, in the highest degree suspicious, particularly when combined with rheumatoid pains in the muscles of the shoulders, breast, and back. It may be that ten, or even twenty years have passed since the infection was contracted, so that patients really no longer think of its being connected with the long forgotten venereal affection; still a trial of iodide of potash, especially in the case of married persons, or ladies of rank, appears to be not only justified, but also directly indicated. If ulcers are present, or affections of the bones, which usually can only be made out by using a probe, one seldom is wrong, in the case of adults, in declaring the case to be one of syphilis. Nevertheless, great caution is necessary, since lupus of the nose, though it occurs but seldom, brings about the same changes, while perforations of the septum may be caused by abscesses, blood cysts, or the use of bichromate of potash, arsenic, or corrosive sublimate. The diagnosis is absolutely certain when other syphilitic symptoms, especially copper-coloured skin eruptions, nodes on the bones, muscular gummata, or cicatrices in the pharynx and larynx, are present.

The prognosis, provided the patients are not too late of coming under treatment, is favourable, although deformity of the nose, stenoses, or cicatrices cannot always be prevented. When the process attacks the facial or cranial bones, death may follow from meningitis or convulsions.

Treatment.—Treatment must be both general and local. In hereditary and early forms of syphilis, mercury must be administered; in the later forms, iodide of potash, which hardly ever fails. Local treatment consists in very careful cleansing with chlorate or permanganate of potash, insufflating iodoform, and boracic acid, and in cauterising the plaques, condylomata, and deeper ulcers with solid nitrate of silver or chloride of zinc.

Schuster and Sänger highly recommend scraping the ulcers and polypoid vegetations with the sharp spoon. When the vegetations are widely spread, I prefer to use the cold, or the galvano-caustic snare. Abscesses must be opened as soon as possible, and necrosed pieces of bone, when not too firmly fixed, removed with forceps or pliers. The chronic atrophic catarrhs, which often remain after cure of the more severe forms, must be treated on the principles already laid down.

TUBERCULOSIS.

Tuberculosis of the nose occurs much less frequently than that of the pharynx.

It is met with both primarily and secondarily, in the form of nodes, and of miliary eruptions. Its favourite situation is the septum. The nodular form may long remain unchanged, until, usually later than the miliary form, it becomes converted into superficial lenticular ulcers of atonic character. Very few communications have been made on tuberculosis of the nose, and these by Laverau, Riedel, Trendelenburg, Weichselbaum, and Tornwaldt. The last named observed, on the anterior end of the inferior turbinated bone, and on the nasal floor, greyish-red non-ulcerated tumours, containing numerous miliary tubercles, with giant cells, and at the same time discovered on the roof of the pharynx, a shallow ulceration, the posterior end of the septum and the larynx being similarly affected. According to Volkmann and Demme, the more serious forms of scrofulous ozæna also depend on miliary tuberculosis. Demme observed, on the septum of scrofulous children, besides greyish-yellow nodes, irregular ulcers, varying in size from a millet seed to a pea, spread out towards the surface, and containing numerous tubercle bacilli.

The symptoms are usually those of chronic nasal catarrh, with fœtid sanguineous secretion.

The diagnosis is founded on the appearance of the ulcers, and on the presence of other phthisical symptoms. According to Weichselbaum, one may very easily mistake them for enlarged mucous glands, or for more pronounced adenoid vegetations and hæmorrhagic erosions. Microscopic examination of the nodes and ulcers here also affords the surest means of diagnosis, by revealing giant cells and tubercle bacilli.

The prognosis is, of course, hopeless. Even when local healing occurs, tubercular deposits appear sooner or later in other organs.

Treatment.—Treatment consists in excision, scraping, or destroying the nodes with the galvanic cautery, in cauterising the ulcers with nitrate of silver or chloride of zinc, and further, in thorough cleanliness, and in the insufflation of iodoform.

SCROFULA, LUPUS, AND GLANDERS.

Up to the present day less is known of scrofula of the nasal cavities than even that of the pharynx.

First of all comes the question—What is understood by scrofula? According as the extent and limits of scrofula are narrowed or widened, the frequency of strumous diseases of the nose will, as Gottstein remarks, be reckoned greater or less. It is a fact that scrofulous children are peculiarly often attacked by acute and chronic nasal catarrh, especially the form accompanied by purulent secretion. In many, fœtid rhinitis is looked on only as a manifestation of scrofula. According to Stoerk, fœtid rhinitis in the scrofulous often goes on to follicular ulceration, which gives rise to ulcers and destruction of the periosteum and perichondrium, and also to perforations of the septum, and caries and necrosis of the spongy bones, or of the vomer.

The destruction does not appear to proceed altogether from the mucous membrane, but sometimes to be brought about by primary disease of the cartilages and bones. On account of the absence of pathological examination, only clinical observation can be depended upon, which by itself cannot be decisive. As far as my own experience goes, I have seen a considerable number of children, without any trace of syphilis, in whom were defects and losses of the septum and turbinated bones, which were no longer active. The majority of them had scars on the neck in consequence of previous lymphadenitis. In one child of twelve, whose septum was wasted away to a small doubly perforated piece of tissue, there were retiform cicatrices in the nasopharynx, cicatrices on the skin, and the lips were thick and swollen, nevertheless, in the absence of any other symptoms in the patient himself, or in his parents, brothers and sisters, I was compelled to accept scrofula as the cause. B. Fränkel, among others, says that although such defects on the septum and turbinated bones are not common, still they undoubtedly exist, and one has no right, particularly in the case of children, to infer that the case is syphilis, solely on the presence of such changes. Since scrofula and tuberculosis are so nearly related, it may perhaps be warrantable to trace such isolated forms to local tuberculosis or to lupus.

Lupus, as a rule, attacks the interior of the nose only when the external nose is already affected. In rare cases the mucous membrane of the nose is primarily and alone affected, sometimes in the form of lupus exedens, sometimes lupus non-exedens. Both forms attack the cartilaginous septum most frequently; the disease appears as little red nodes, of the size of a pin's head, which after a time become changed into ulcers with fœtid secretion, spreading both superficially and in depth, and result in perforations of the septum or necrosis of the alæ nasi. Necrosis of the nasal bones, according to Kaposi, does not occur in lupus. In the non-ulcerating form, retraction of the mucous membrane and of the underlying cartilages and bones takes place.

The diagnosis is sometimes very difficult, particularly when symptoms of syphilis are also present. By taking into account the youth of the patient, his often very pronounced scrofulous habit, the slow spread of the disease, the inefficiency of iodide of potash, and the participation of the skin and lymphatic glands, a certain diagnosis may always be come to.

The prognosis is so far doubtful, in that after spontaneous or artificial cure, great tendency to relapse exists, and other organs may also become affected.

Treatment.—Treatment consists in destroying as early as possible the affected parts, either with the galvanic cautery or with chloride of zinc, in careful cleansing of the nose, and in improving the constitution by means of iron, arsenic, or cod-liver oil.

Glanders is a disease which is communicated to men by animals, and chiefly by the horse. Besides the mouth and pharynx, it affects the nose also, in quite a special way.

It principally attacks grooms, coachmen, veterinary surgeons, pathological anatomists, in short, all persons who have to do with infected animals. On the inflamed mucous membrane of the nose there develop small yellow nodes, which disintegrate, leaving behind ulcers of greater or less size, and giving rise to necrosis of the cartilages and bones. According to Bollinger, the affection of the nose, both in men and in horses, often first appears in the second or third week, at the conclusion of the disease, and even then it is by no means pathognomonic.

The symptoms at the beginning consist in the secretion of a thin tough mucus, followed gradually by swelling, redness, and great tenderness of the nose and surrounding parts, especially the root of the nose, which often becomes thickened, sensitive, and diffusely swollen as in erysipelas. The discharge from the nose, which is often from one side only, later on becomes thicker, more purulent, brownish yellow, sanguinolent, and fœtid. One is seldom able to discover real nodes, which most often occur on the alæ nasi. The more serious cases go on to perforation of the septum, caries and necrosis of the vomer, considerable swelling of the submaxillary glands, and may end in death from pyæmia.

The diagnosis is all the more difficult because of the want of any etiological factors, and because the ulcers bear the strongest resemblance to those of syphilis and other affections. The febrile commencement, the almost constant pain in the joints, the appearance of glanders pustules on the face, the inefficiency of iodide of potash, and the rigors and shivering, put one, as a rule, on the right track.

The prognosis of acute glanders is absolutely unfavourable. It is more favourable in the chronic form, nevertheless marasmus may remain for the rest of life.

Treatment.—Treatment consists in cauterising the diseased mucous membrane with nitric acid, chloride of zinc, or carbolic acid, also in syringing with a solution of creasote (1 in 5000) several times a day. It is very evident, that in so terrible a disease, which can be transmitted from one person to another, the greatest cleanliness is necessary, and that all linen, &c., which have become dirtied by the secretion should be destroyed.

PARASITIC DISEASES.

Amongst the parasitic affections occurring in the nose, *soor or thrush* is relatively the most frequent. The occurrence of *oïdium albicans* on mucous membranes with ciliated epithelium is quite exceptional. Thrush occurs in the nose almost only when it is also present in the mouth or pharynx, and then chiefly in children, rarely in adults. It excites irritation of the mucous membrane, with profuse secretion and obstruction of the nose.

About other fungi little is known. We only know that numerous kinds of mould, and of bacteria, occur in the fœtid secretion of the various nasal affections, and some observers consider that these bacteria alone are often the cause of the bad smell in so many nasal secretions.

Macroscopic parasites of the nose must be looked upon rather as foreign bodies, since there are no animals to which the human nose is assigned as a breeding place. Such parasites usually enter the nose during sleep or intoxication, especially if it contain fœtid secretion. The ascaris lumbricoides somewhat frequently, and earwigs, centipedes, threadworms (oxyuris vermicularis), maggots (œstrus), the pentastoma tænioides, flies' eggs, and leeches have also been met with. This affection occurs far more frequently in the tropics than with us; the Indian "Peenash" is very often due to the immigration of the eggs of the lucilia hominivora.

The symptoms, which these parasites excite, are due to the kind and size of the insect. The first symptom is itching and sneezing, then the affected nostril becomes stopped up, and a mucous, sanguineous, or purulent secretion and epistaxis set in, or there are œdematous swelling of the face and fever, throbbing pain in the head, giddiness, sleeplessness, cerebral irritation, convulsions, and coma. In rare cases, patients have a feeling of some object moving in the nose. B. Fränkel in his book relates that when the French army was in Mexico, numerous suicides took place in consequence of the frightful pain excited by the larvæ of insects in the nose.

The diagnosis of thrush can only be made with the microscope, otherwise it is impossible to help mistaking it for diphtheria. The diagnosis of parasites is sometimes extremely difficult, particularly when those attacked can give no clue, but if very careful and oft repeated inspection be made, the trouble cannot long remain concealed.

Treatment.—Treatment must be directed to removing the parasites.

An attempt should always be made to drive them out of their hiding-places with the nasal douche, before proceeding to extract them with instruments. To kill the parasites, inhalations of ether, alcohol, benzoin, chloroform, sal ammoniac, and turpentine, fluid injections containing alcohol, alum, creasote, carbolic acid, and chloroform, infusions of tobacco, and oak bark, insufflations of iodoform, carbolic, or thymol powder, calomel, and corrosive sublimate are all considered good. Mankiewicz recommended painting with balsam of Peru. As a rule, all symptoms disappear after the parasites are removed. If any discharge from the nose remain, it must be treated on the rules already mentioned.

The treatment of nasal thrush is the same as that of the mouth.

HÆMORRHAGES.

No organ shows so great a tendency to hæmorrhage as the nose. The cause lies partly in its exposed position, and partly in its extraordinary richness in blood vessels.

Amongst the exciting causes, the most frequent are wounds, falls, or blows on the nose, fractures of the bones and cartilages, entrance of foreign bodies, scratching and picking with the fingers, violent blowing the nose, introduction of a probe, catheterising the Eustachian tube, and other operative interference. Besides these may be added, spontaneous symptomatic hæmorrhages in the course of acute and chronic catarrhs, ulcers, plethoric neoplasms, foreign bodies, and parasites. Hæmorrhages from engorgement and active hyperæmia occur during whooping-cough, and diseases of the lungs and heart, particularly in aortic insufficiency and general plethora, and also in some liver and spleen affections. The acute infectious diseases, especially scarlatina, measles, typhoid, recurrent fever, and diphtheria, supply a considerable contingent. Habitual or often recurring hæmorrhage without disease of the nasal mucous membrane, is quite common in anomalies of the blood, chlorosis, anæmia, pregnancy, hæmophilia, morbus maculosus, leucæmia, and scorbutus. Here also may be mentioned hæmorrhage occurring during puberty, and from hereditary plethora. The next cause lies in abnormal fragility of the vascular walls. Comparatively still less known causes of habitual epistaxis are dilated small vessels, pin head telangiectases, and erosions of the septum. These are most often met with on the most antero-inferior part of the cartilaginous septum, from one to one and a half centimetres above the level of the floor of the nose, and seldomer on the middle of the septum, or on the anterior ends of the spongy bones. Of the occurrence also of vicarious epistaxis there is no doubt. It takes the place of hæmorrhage from the anus, as in hæmorrhoids, or of menstruation. Several cases of epistaxis, occurring regularly every four or six weeks, in place of the menses, and disappearing with pregnancy, are cited by B. Fränkel. Intermitting hæmorrhages from the nose occur in intermittent fever.

Of much less frequent occurrence than free hæmorrhages, are those into and under the mucous membrane. They appear either as little pin-head or bean-sized ecchymoses, or as tumour-like swellings—hæmatomata. They arise either spontaneously or from injuries, particularly after fractures of the bony and cartilaginous septum. Their surface is smooth and soft, their colour from deep red to black, and the surrounding mucous membrane appears ecchymosed. Usually the swelling is bilateral, and fluctuating, and it may be so large as to protrude from the nostril.

In free hæmorrhages the blood flows either gradually and slowly, or quickly and in a stream, from one or both nostrils. When the hæmorrhage is from the posterior part of the nose, the blood flows through the nasal fossa and is swallowed, vomited, or expelled through the mouth. The quantity of blood lost is extremely variable, —from a few drops to several pounds. The colour is generally bright red, but when coagulated or slowly effused it becomes brownish-red or black. Some forms of nasal hæmorrhage are preceded by giddiness and flushing of the head, dull pressure, throbbing in the forehead, tinnitus, and sleeplessness, afterwards the patients generally feel very much relieved. Habitual epistaxis often leads to anæmia and weakness, very violent acute bleedings may give rise to fainting, weakness of the heart, and even to death. In profuse hæmorrhage during sleep or fainting, the danger of inspiring the blood is very great. Ecchymoses usually occur without symptoms, larger hæmatomata cause obstruction of the nose, interference with smell, and in some cases dull pain.

The diagnosis of free hæmorrhage is generally easy, especially if the blood flows from the nostrils, or if the bleeding spot can be seen. The diagnosis is more difficult when the blood flows through the fossæ, or when it is doubtful whether the blood comes from the lungs, pharynx, or stomach. Hæmoptysis is usually preceded by extensive dull pain in the chest, or by sharp pain, cough, and other symptoms of phthisis, or of valvular lesions ; hæmatemesis, by stomach symptoms, such as vomiting, nausea, and eructation. Hæmorrhages from the throat are recognised on inspection, or on employing posterior rhinoscopy, hæmatomata and ecchymoses by anterior rhinoscopy.

The prognosis depends on the cause and duration or severity of the bleeding. It is very unfavourable in fractures of the skull and in cases of hæmorrhagic diathesis, because in cases of the latter the hæmorrhage recurs, and we possess no remedy to cure the fundamental disease. It is self-evident that profuse hæmorrhages with fainting, disappearance of the pulse, &c., are always dangerous, especially if skilled help be not to be had, and also that profuse, or recurring bleedings in the course of severe febrile diseases, of typhoid, diphtheria, &c., present, to say the least, a very doubtful prognosis.

Treatment.—As some hæmorrhages, especially those occurring in persons affected with plethora, hæmorrhoids, heart disease, and amenorrhœa, exercise a beneficial influence on the general condition, no treatment is necessary so long as the bleeding is not excessive. In other forms, treatment must be regulated by the intensity of the hæmorrhage. When it arises from the anterior part of the septum, it can most easily be controlled by compressing the corresponding ala of the nose with the finger, or by introducing a piece of wadding, sponge, &c. Snuffling, blowing the nose, and bending forward the head, as are so often done, prolong the bleeding, and must be absolutely forbidden. In moderate bleeding from the deeper parts of the nose, the nasal douche is recommended, or the snuffing up of some astringent solution containing alum, vinegar, or liq. ferri perchlorid. (15 drops to a pint of water), but when the bleeding is more profuse, these and the numerous other popular remedies are useless.

The surest method is tamponading or plugging. In most cases

plugging anteriorly is enough. A piece of ordinary, or, better, of hæmostatic wadding, about a finger's length, and one to two centimetres thick, rolled round with a thread, is introduced into the nose by means of a forceps or of Gottstein's screw, and firmly pushed into the upper parts with a probe, then a second piece, and if necessary a third, till the nose is filled; the threads hanging out of the nostrils are either tied round the ear, or fastened to the bridge of the nose with a piece of plaster. After twenty-four hours the plugs are removed.

If this does not succeed, then plugging the posterior nares must be reverted to.

In cases of necessity an elastic catheter does very well. Through its apertures a piece of thick waxed thread, catgut, or strong antiseptic silk, is passed; the catheter is now introduced into the nostril and pushed back till the thread is seen behind the soft palate, when it is seized with forceps and drawn out of the mouth, and a dry piece of wadding, about the size of a walnut, is tied to it several centimetres from the end. By drawing back the catheter, the tampon, with the help of the forefinger, is brought into the naso-pharynx and then pushed into the nasal fossæ. The short thread lying in the pharynx serves to draw out the plug, the longer one hanging out of the nose is to be fastened in the way mentioned above.

Plugging is easier when Bellocq's sound is used, as the spring with the thread fastened to its anterior end passes more quickly round the palate; only it would be desirable that the instrument to be used in future were more slender and neat. The introduction and fastening of the plug takes place in the same way as with the catheter. The plug should not be left in the nassal fossæ more than from forty-eight to sixty hours, as gangrene, tetanus, erysipelas, and pyæmia have occurred by its being left in too long.

Among the numerous instruments for this purpose, the rhineurynter or rhinobyon of St Ange must be mentioned, and also the intra-nasal plug of Cooper Rose. Each consists of a thin indiarubber bag, which is connected with a tube provided with a stop-cock. The bag is passed empty along the floor of the nose into the nasal fossæ, and then blown up or filled with water.

If the bleeding still continue, or there is danger in delaying, then transfusion is indicated.

The main indication for treatment is, in the second place, to remove the fundamental disease, and prevent the recurrence of the hæmorrhage. Such a radical cure is almost always possible in those cases of bleeding, which owe their origin to pathological changes of mucous membrane, e.g., chronic catarrh, polypi, foreign bodies, parasites, ulcerations, and excoriations. It is very often possible to improve the blood, and lessen the predisposition to epistaxis, by the inward administration of such remedies as iron and quinine, or by residence in the country. The most suitable styptics to be taken inwardly are ergotin (which may also be administered subcutaneously), liq. fer. perchlor., sulphuric acid, and acetate of lead. Plethoric persons, and those subject to hæmorrhoids, should be sent to Karlsbad, Kissingen, Homburg, or Marienbad; or, the former should be purged at home for a considerable time, and most strictly enjoined to avoid alcohol, while for the latter, prophylactic blood-letting by means of leeches to the anus is recommended. In order to bring back menstruation to its normal condition, scarification of the vaginal portion of the uterus, hot douching, hot foot baths, and iron and aloes pills should be prescribed.

Hæmatomata must be emptied as soon as possible from their most dependent part, or, if necessary, split through their whole extent, just as if their contents had already become purulent.

FOREIGN BODIES AND CONCRETIONS.

Foreign bodies in the nose occur almost only in children, very rarely in adults, and then generally in the insane.

Among the ordinary things that children put into the nose while at play, are peas, beans, buttons, cherry and plum stones, stones, pieces of wood, small sponges, paper and other pellets, bits of shoe leather, &c. In adults there sometimes remain, after injuries, bullets or broken off knife blades, or other objects; as, for example, in a case of Voltolini's, where the patient, after a fall, carried a large number of broken reeds in his nose for three months. Plugs also may be left, through forgetfulness, and pieces of food may become lodged in the nasal fossæ during the act of vomiting.

The symptoms, which foreign bodies give rise to, are caused by their physical properties, and by the length of time they remain in the nose. Beans and peas sprout, and throw out roots and outgrowths. Pointed objects, soon after their entrance, give rise to bleeding, as well as pain; rounded objects excite swelling of the mucous membrane, with at first serous, afterwards purulent secretion, and render the affected side of the nose more or less impermeable. The longer a foreign body remains in the nose, the more fœtid becomes the secretion, the more violent the headache and pressure in the brain. The ordinary sequelæ are ulceration, necrosis, and polypoid vegetations. The skin of the nose and cheeks often swells up, and the upper lip is eroded by the pungent fœtid sanguinolent secretion, while displacement of the naso-lachrymal canal gives rise to epiphora, conjunctivitis, and other affections.

Similar symptoms are set up by concretions. Small foreign bodies generally give rise to their formation, by becoming superficially encrusted with lime. Sometimes the nucleus of these concretions or *rhinoliths*, is formed by dried-up secretion, or by a coagulum of blood. In a case related by Schmiegelow, nothing of this kind could be made out. Hering once observed the nasal cavities almost completely closed by a hard substance, like cement. The main part of these rhinoliths, or about 80 per cent., consists of inorganic substances, chiefly phosphate and carbonate of lime, a little oxide of iron, and about 20 per cent. of organic material.

According to B. Fränkel, this stone formation is not to be confounded with calcification of the mucous membrane; the latter occurs in the noses of old people, sometimes also in younger persons, in consequence of the diathesis ossificans. Particularly the mucous membrane of the accessory cavities, and of the spongy bones, may calcify in the form of fine grains and laminæ, and may assume a white appearance.

The diagnosis of foreign bodies and concretions, in the case of children, and where no history can be got, can only be made by direct examination. Fœtid discharge in children should always make one suspect that it is caused by a foreign body. I have rarely been deceived in this, and was once able in a few minutes to remove an affection that had remained unrecognised for years, and had been treated with all sorts of lotions. On account of the great opposition children make, an anæsthetic is indispensable for an exact examination of the nose, narrow enough at any time. As inspection, often on account of the enormous swelling and copious secretion, gives no, or only very insufficient information, the diagnosis comes to depend, for the most part, on probing; this, however, may also mislead, as it may, for example, cause one to mistake a rhinolith for a sequestrum of bone.

Treatment.—The removal of foreign bodies requires not only caution, but also considerable skill. Roundish bodies recently put into the nose may, under favourable circumstances, be removed by blowing the nose, or by the introduction of snuff, or of the nasal douche into the free nostril. Objects which have been in some time, and have been driven further back, must be extracted under the guidance of the mirror, by means of forceps, pliers, or other suitable instruments. I myself prefer to use, especially in the case of smooth objects, which may so easily be pushed deeper into the nose, a probe, spoon-shaped, or fashioned into a loop at its extremity, which is cautiously brought behind the foreign body. In children anæsthesia is generally indispensable, but quite unnecessary in adults.

Concretions, which, on account of their size, or their hardness, cannot be extracted *in toto*, must first be lessened by means of forceps, or of lithotrite-like instruments, and then removed piecemeal. To push them directly through the fossæ into the pharynx, as may be done with some foreign bodies, is rough, and is only justifiable when removal through the anterior nares is impossible. During the manipulation, either breathing must be forbidden, or the finger must be introduced into the naso-pharynx.

NEOPLASMS AND TUMOURS.

It is very difficult to draw a hard and fast line between hypertrophies of the mucous membrane and real neoplasms. It will be well, therefore, considering the numerous transition forms, which have already been mentioned under chronic rhinitis, that, after Hopmann's example, we should designate the latter, *i.e.*, the cavernous and telangiectatic tumours and papillary formations, polypoidal tumours, while the real neoplasms should be known simply as polypi.

The etiology of these swellings is somewhat obscure. That irritable conditions of the mucous membrane, and particularly chronic catarrh, figure to a certain extent, or even at all in their causation, appears to me very doubtful, from the fact that mucous polypi in one nostril, and hypertrophy of the mucous membrane in the other, or both in the same nostril, at the same time, is quite a common occurrence. Still the fact remains that, in spite of the great frequency of nasal catarrhs in children, neoplasms are observed only in exceptional cases, before the age of fifteen. Of the etiology of malignant neoplasms of the nose, as of other parts of the body, very little is known.

The commonest form of non-malignant neoplasms are *mucous polypi*.

They are seen as soft, œdematous, jelly-like, transparent tumours, bluish-white or yellowish, more seldom reddish or deep red in colour, and varying in size from that of a pin's head to that of the thumb. They are usually oval in form, or pear-shaped; the larger ones always assume the form of the space in which they grow, and are, therefore, generally flat, and longer than they are broad. If they have peduncles, they are suspended in the nares as longish sack-like formations, movable with every breath, but if their peduncle is very wide, or their insertion diffuse, they are seen as irregular, slightly moving, or immovable tumours. They most usually spring from the edge of the middle turbinated bone, from the nasal roof, from the outer wall of the nose, or from the space between the inferior and middle turbinated bones, more rarely, however, from the former, and still more rarely from the septum, or from the floor of the nasal cavity. They are always in groups, one by itself being a very great rarity. There are usually several present, sometimes as many as twenty, or even thirty to forty. Once I removed as many as sixty-five from the nostrils of a patient. Generally both nostrils are affected, rarely one. In well marked cases both are completely obstructed from front to back. Mucous polypi contain the elements of the mucous membrane from which they arise, their epithelium almost always consists of ciliated cells, the movement of which, in a recently extirpated polypus, affords a most interesting object for demonstration.

According to Billroth, sometimes the connective tissue, sometimes the glandular tissue preponderates. In the former case they appear like soft fibromata, in the latter like adenomata, whose newly-formed hypertrophied glands are held together by a widely cancellated œdematous connective tissue. Hopmann designates mucous polypi as soft fibromata, consisting of a widely meshed network of areolar connective tissue, the meshes of which are filled partly with cells and partly with serum-albumen, and which sometimes contain serous or purulent cysts.

Another form of nasal polypi are the *papillomata*. According to Hopmann, whose assertion I can uphold, they occur exclusively on the inferior turbinated bone, as partial or diffused broad based tumours of papillary structure, rarely pedunculated or sharply defined from the healthy mucous membrane. The papillæ are either closely packed together or loosely separate, in the latter case the individual papillæ are more developed, and are connected like berries on short thick stalks. Their colour varies from bright rose to a cherry red, their consistence from the softness of a mucous polypus to the hardness of a fibro-sarcoma. They are mostly soft, berry-like, more rarely composed of papillæ arranged like laminæ. Other observers, among them Schäffer and Zuckerkandl, describe these papillomata as telangiectatic tumours.

Tumours, consisting for the most part of compact connective tissue, are designated *fibromata*. These occur very rarely in the anterior parts of the nose, they are more frequently met with in the posterior divisions of the nose, and there represent naso-pharyngeal polypi. Besides connective tissue they generally contain elastic fibres, round cells, and dilated vessels, and may, therefore, also be called fibro-sarcomata. They arise almost exclusively from the superior wall of the nose, from the periosteum of the vomer, and sphenoid, from the sphenoidal fissure, or from the aponeurosis of the foramen lacerum anterius. They are seen as compact globular or bottle-shaped tumours, generally with broad pedicles and of slow growth, but often attaining an enormous size, filling the entire pharynx, and growing forward into the nose. The more sarcomatous forms, which are very rich in cells, more usually penetrate the cranial cavity, while the purely fibrous must be considered as benign tumours.

The *sarcomata* belong to the malignant neoplasms, consisting of small or large, round or spindle cells. They are usually soft and very rich in blood. Carcinomata and epitheliomata happily occur but very rarely. They consist of pavement epithelium, and are seen as soft cauliflower-like papillomatous growths, with capillary loops, which grow with extraordinary rapidity, penetrate all fissures, perforate neighbouring bones, seize upon the external skin, and extend rapidly into the orbital, oral, and cranial cavities.

The crests and spines on the septum, previously referred to, are partly enchondroses, partly exostoses, and hyperostoses. Besides these, there also occur echinococcus tumours, enchondromata, osteomata, myxomata, ivory exostoses, and herniæ cerebri.

Amongst the most remarkable formations are those tumours which partially or altogether consist of hairs, or which contain cartilage, bone, fat, connective tissue, and glandular substance. The formation of teeth in the nose has several times been observed, and among others, by Schäffer and Fletcher Ingals.

The symptoms of neoplasms in the nose are caused by their size and situation. Smaller tumours, especially mucous polypi on the nasal roof and on the middle turbinated bone, may run their course utterly without symptoms, and are often only discovered by chance. As they continue to grow, the already mentioned symptoms of chronic rhinitis with nasal obstruction make their appearance. Telangiectatic tumours, particularly when on the septum, and also papillomata, give rise to frequent and profuse hæmorrhage. From the nose becoming filled up with very many or very large polypi, not only do the spongy bones become atrophied, and the septum displaced, but the external appearance of the nose also becomes altered, the bridge looks broader and flatter, and the nasal bones form an obtuse angle. In some cases there is a feeling as of some foreign body moving about in the nose, and often the polypi protrude from the nostrils. Naso-pharyngeal polypi give rise not only to alteration in speech, but also to difficulties in swallowing, or to the sensation of a foreign body in the pharynx, to tendency to vomit, copious secretion, dull pressure in the head, often to quite remarkable sleepiness, and also to disturbances of hearing. From pressure on the naso-lachrymal canal epiphora may arise; by closure of the sinus apertures, hydrops of the antrum of Highmore, or empyema of the frontal sinuses, may be set up. Malignant neoplasms have, at first, the same symptoms. With their growth there result purulent, often very fœtid, secretion, violent and frequent epistaxis, pains in the nose and under the eyes, exophthalmos, redness and ulceration of the skin, cachexia, pyæmia, and cerebral symptoms.

In the case of all neoplasms, symptoms on the part of the nervous system occur particularly often, such as headache, giddiness, failure of memory, inability for mental work, and also the previously mentioned reflex neuroses. Of these last, however, only asthma and its relation to nasal polypi will here be discussed. And first of all, the fact must be affirmed, that asthma is comparatively seldom observed, considering the extraordinary frequency of polypi. All observers are also agreed on this, that asthma occurs far more frequently in cases where polypi do not entirely obstruct the nose, than in cases where both nostrils are impermeable. The reason for this is found, so far as can be gathered from observations hitherto made, in the excitability of the nerves of the mucous membrane, particularly of those of the erectile tissue of the inferior turbinated bone. By the irritation which the polypi exercise upon the nerves of the mucous membrane, there arises an engorgement of the erectile tissue, which either immediately or by gradual carbonic acid intoxication causes asthma. The occurrence of the attacks during sleep, and their absence in those patients with polypi, who must sleep with their mouths open, mainly support the theory of carbonic acid intoxication in consequence of insufficient nasal respiration. The absence of asthma, in complete obstruction of the nose from polypi, is explained on the one hand, by the lowering of the sensibility of the sensitive mucous membrane from being in constant contact with the neoplasm, and on the other, by the impossibility for stimulus from without to reach the mucous membrane.

The diagnosis must be founded on inspection and palpation. While the former affords information more generally as to the presence of polypi, the probe gives more precise instruction as to the manner of attachment, size, and number. Pedunculated polypi move backwards and forwards with respiration, while larger and sessile ones appear altogether immobile. The diagnosis of smaller tumours on the roof of the nose, or deeply situated in the middle meatus, is very difficult, even impossible, when deviations of the septum or spines are also present. Naso-pharyngeal polypi can only be made out by posterior rhinoscopy and digital examination.

The prognosis must depend on the situation and on the histological structure. No one can deny, that in consequence of the advances of nasal surgery, the prognosis is now much more favourable than formerly. Restoration of nasal respiration can almost always be accomplished, and with it removal of the symptoms caused by it. The prognosis is obscured only to a certain extent by the decided tendency of most polypi to return, hence a radical cure is very often effected. Recurrence is frequently due to too little perseverance on the part of the patient, but also to insufficient treatment of the pedicle, and the impossibility of getting at polypi deeply concealed. Even sarcomatous polypi may be thoroughly cured, when favourably situated. Of course, malignant tumours, especially soft carcinomata, give a hopeless prognosis.

Treatment.—Treatment can only be operative. Cauterising with lunar caustic, touching with chromic acid, or electrolysis, can only be looked on as tedious and purposeless. It is quite evident that the method formerly, and still at the present day sometimes, used, viz., the tearing out with forceps, only exceptionally accomplishes a radical cure; and of what use is it to the patients usually afflicted with numerous polypi, if only the most anterior and most visible polypus is partly or altogether removed? Apart from the uselessness of the method, it is, as Voltolini, years ago, rightly said, a very rough and painful one, all the more as injuries to the turbinated bones, and tearing away healthy parts of the mucous membrane, are quite unavoidable.

The method least hurtful to the patient, is removal by means of the cold snare, for the introduction of which credit is chiefly due to Zaufal. Those who have once been operated on, by means of forceps, can never sufficiently praise this method. I have modified Tobold's wire snare by strengthening the handle, and placing a ring on the posterior end, which, by being able to rotate, gives to the instrument greater handiness and strength for operations in the nose. The tube, through which good steel or brass wire is drawn, must be at an obtuse angle to the handle, and at least 14 centimetres long, and must have a cross-bar about a millimetre from the anterior end, against which the tumour, when seized, can be cut through. The wire being pushed through, is fastened by winding it several times round one of the hooks fixed into the sliding part.

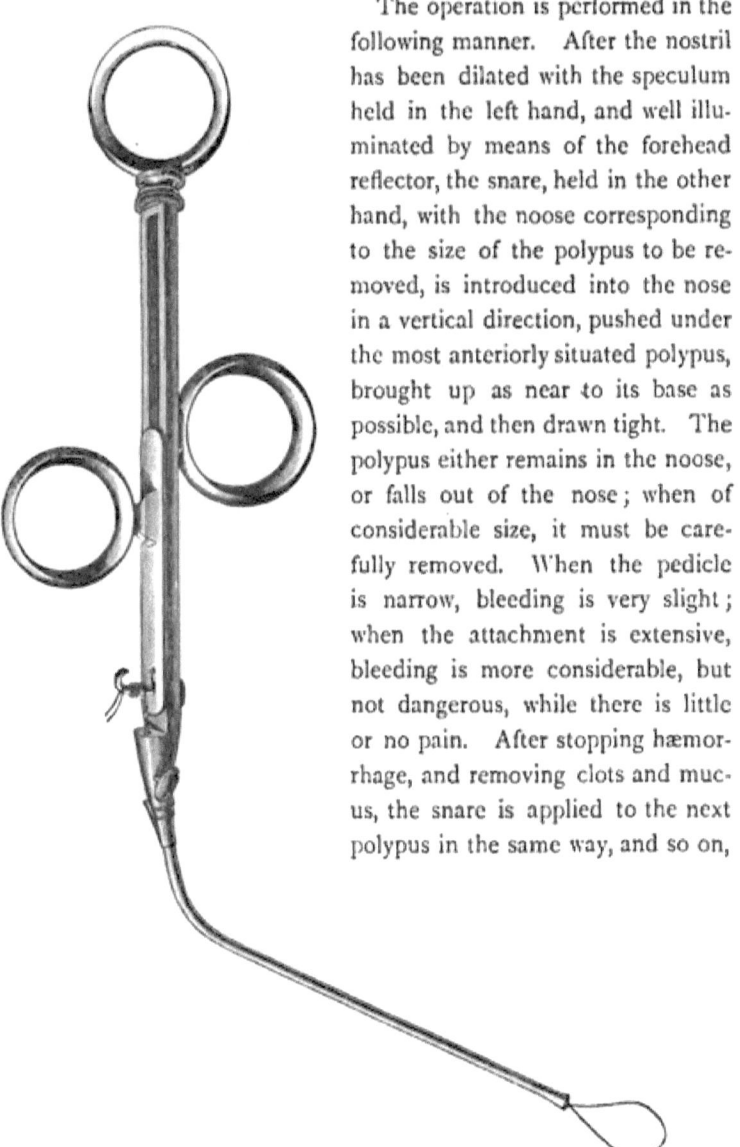

The operation is performed in the following manner. After the nostril has been dilated with the speculum held in the left hand, and well illuminated by means of the forehead reflector, the snare, held in the other hand, with the noose corresponding to the size of the polypus to be removed, is introduced into the nose in a vertical direction, pushed under the most anteriorly situated polypus, brought up as near to its base as possible, and then drawn tight. The polypus either remains in the noose, or falls out of the nose; when of considerable size, it must be carefully removed. When the pedicle is narrow, bleeding is very slight; when the attachment is extensive, bleeding is more considerable, but not dangerous, while there is little or no pain. After stopping hæmorrhage, and removing clots and mucus, the snare is applied to the next polypus in the same way, and so on,

Fig. 5.
Wire snare for the nose (cold snare).

till all are removed. When pedicles are thin, bleeding little, and polypi few, the operation is completed in a very short time.

It is quite another thing, however, when both nares are completely impermeable, or when one has to do with extensively attached flat growths. In the former case, the trickling, and even flowing blood, and the mucus which is copiously secreted in consequence of irritation of the mucous membrane, must, after each operation, be removed with forceps or a syringe, which not only causes great delay, but which is often very difficult. If the nose is entirely filled up with polypi, then, as I have often observed, repeated operations are necessary before respiration through the nose can be restored.

Polypi with thick pedicles, but still more those of very flat growth, offer considerable opposition to the use of the cold snare. It may happen that, owing to its unexpected thickness, a pedicle may not be cut through. The perplexed beginner, in this trying situation, requires only to cut the wires through at their posterior end with scissors, and draw back the snare. The noose holding the polypus can easily be removed by rotating and pulling upon one end of the wire. Polypi, when extensively attached, must therefore be removed with the galvano-caustic or electric snare; but one must often work with both snares alternately. On account of the high price of platinum, steel wire is recommended for the electric snare; the strength of the current, however, must always be tested beforehand, in case the wire should fuse. The greatest difficulties are met with in removing neoplasms from narrow or crooked noses. But, with practice and skill, these hindrances will also, in time, be overcome, by learning to make use of the most limited space.

As cutting away what is above ground is not enough for the extirpation of a weed, similarly one must not be content with simply cutting away polypi.

Here begins the most important part of the treatment, the destroying the stump and the place of origin. The more energetically and carefully this is done, the more favourable are the prospects of radical cure; the more carelessly, the more certain are the polypi to recur. It is often most difficult to convince patients of the necessity of this procedure, they being quite content with being again able to breathe through the nose. Although cauterising with solid caustic, chromic acid, or chloride of zinc, is not to be rejected, still the galvanic cautery is far the best remedy. Suitably formed cauteries are pressed against the stump of the polypus, or are made to penetrate it, and destroy as much of it as possible. The cauterisation is all but painless, so long as diseased tissue only is touched, becoming more painful, however, the nearer the healthy tissue is approached; shooting pains in the teeth, too, are generally complained of.

If the operation is conducted in this way, radical cure may take place in cases where it seemed scarcely possible. Not so favourable, however, are the prospects in sarcomatous and in extensively attached polypi. Although they remain apparently cured for one or several months, or even for a year, still, in spite of the galvanic cautery, they may again recur. In such cases, M. Mackenzie recommends that, not only a piece of the mucous membrane, but the entire spongy bone be removed with forceps, or with a cutting chisel.

Considering matters from a practical point of view, it may not be out of place to notice here some of the unpleasant incidents, reaction and after-treatment of these cases.

The most frequent of these disagreeable events, during the operation, is hæmorrhage. It has been already mentioned that the galvanic cautery cannot be too highly praised, in that hæmorrhage is, by its use, almost entirely prevented. Although this is not altogether the case, still it is true that profuse hæmorrhage can be obviated by its use, by frequently interrupting the current, and thus preventing the instrument getting to white heat. Insignificant hæmorrhages can generally be stopped by injecting cold water, or a solution of perchloride of iron, and when more profuse, not by plugging with Bellocq's sound, but by pressing a piece of wadding, or a piece of fine sponge, impregnated with perchloride of iron solution, against the bleeding point.

Sometimes there arise, from irritation of the sensory nerves, particulary in operations on the nasal roof and middle spongy bone, attacks of sneezing, or of lachrymation, also nausea, vomiting, and fainting, or a feeling of weakness. Suspending the operation, and applying restoratives, such as smelling salts and wine, will soon cause the patient's recovery. If too many polypi have not been removed, patients may immediately return to their business, or go home; it is best, however, to tell them, that after a time a great deal of mucous, bloody, and even purulent secretion, will come from the nose, and perhaps even stop it up again.

This last condition generally occurs after cauterising a great many polypus stumps, or a large extent of degenerated membrane. A true fibrinous exudation is formed, which takes up space, and is often not thrown off for about fourteen days. A further extension of this croupous exudation, I have, as yet, never observed, and just as seldom the appearance of erysipelas. In some cases reaction is more violent, and patients get slight fever, headache, lassitude, and loss of appetite, and are even compelled to take to bed. By administering quinine, cooling drinks, poultices and a laxative, recovery is soon brought about.

The proper after-treatment consists in washing out the nose two or three times a day with some disinfectant, such as chlorate of potash or boracic acid. This not only succeeds in removing the copious secretion, but also in preventing its putrefaction or infection with diphtheritic poison.

The removal of naso-pharyngeal polypi is one of the most difficult of operations.

The operation may be performed either from the nose, or from the pharynx. If the method through the nose is chosen, the cold and the electric snare next come into consideration. The difficulty is to get the noose to the very base of the tumour; this can be done in two ways, either by turning and raising the noose till it is accomplished, or by pushing the wire introduced into the nasal fossæ over the tumour, with the help of the index finger in the naso-pharynx, which, however, is easier said than done. Zaufal constructed a peculiar wire snare with spring power, by which the loop is formed after introducing the point of the instrument into the pharynx. Because of the extraordinary hardness and extensive attachment of naso-pharyngeal polypi, the electric snare must be preferred on account of its certainty.

It often happens that all attempts fail to secure the tumour in the snare. In such cases I renounce all further experiments, and push directly against the body of the root of the polypus, provided it can be seen, or at least felt with a probe. I have often succeeded, by boring in with a pointed, or knife shaped cautery, in making the base of the tumour slough, so that it shrivels up and falls off. A further advantage of this method is that special after-treatment of the stump becomes unnecessary. If this also should fail, then an attempt must be made to remove it from the pharynx.

Crushing with cutting forceps, I do not consider to be by any means so dangerous and rough an operation as tearing out with ordinary dressing forceps; the difficulty lies only in getting the branches of the instrument sufficiently near the place of insertion; cauterisation with lunar caustic or with the galvanic cautery must also be undertaken here.

The application of the cold and of the electric snares from the pharynx, offers no less difficulty than that from the nose. Although it is desirable to operate under the guidance of the mirror, still it appears to me absolutely useless, especially as want of room so often demands modifications in this respect. Gradual destruction by pressure of a solid galvanic cautery, or of Lincoln's concealed electrode, is certainly very tedious, although no doubt successful.

Malignant neoplasms may for a time be delayed in their growth, though not, of course, cured, by energetic and frequent application of the galvanic cautery. At a later stage their extirpation by means of resection is indicated, though their recurrence even then is all but inevitable. Enchondromata may be removed with the cold snare, as a case related by M. Mackenzie proves, but the electric snare seems more certain. Soft spongy osteomata or exostoses may be broken up with pliers and removed piecemeal; compact ones require chiselling out, or the application of the chain saw, or, if necessary, drilling out after splitting the nose.

Although not always possible, yet, in the majority of cases, neoplasms situated in the nose or in the naso-pharynx may be got rid of by endo-nasal or endo-pharyngeal means. No doubt, therefore, in time, these hitherto so much preferred severe surgical operations, splitting the hard and soft palates, resection of the upper jaw, of the nose, &c., will be reduced to a minimum, all the more, as by means of them no better guarantee against recurrence is given. Fortunately these methods are abhorred by the public, who are now beginning more and more, to prefer the special methods of treatment.

DISEASES OF THE NERVES.

Corresponding to the two principal functions of the nose, sensibility and smell, we differentiate disturbances of the trigeminus or fifth nerve and of the olfactory nerve.

Disturbances in the sphere of the trigeminus are of very rare occurrence.

Anæsthesia of the nasal mucous membrane alone is extremely rare. As a rule, it is caused by central diseases, and paralysis of the trunk of the trigeminus in consequence of cerebral tumours and syphilis of the brain. Insensibility of the nasal mucous membrane is manifested by absence of reaction to touch, and by the absence of sneezing. If at the same time smell appears diseased, the reason is, that the perception of sharp, pungent odours, such as sal ammoniac and acetic acid, is not a sensation of smell, but rather one of taste.

Hyperæsthesia of the nasal mucous membrane, on the other hand, is of pretty common occurrence. It manifests itself by the occurrence not only of unpleasant itching sensations, but also the most severe reflex symptoms, and especially fits of sneezing, whenever irritation, of however insignificant a character, affects the mucous membrane of the nose, the eye, or even distantly situated organs, such as the sexual apparatus and the rectum. The itching feeling in the nose caused by entozoa, especially the ascaris or the tapeworm, is well known. Hay fever and nervous catarrh may also be looked upon as similar neuroses.

Neuralgia of the nasal branches of the trigeminus often occurs before or during the course of an acute catarrh. The dull headache, especially in the forehead, the piercing, beating, and hammering pains synchronous with the pulse, deep in the eyes, in cases of acute and chronic hyperplastic catarrh, of polypi, and diseases of the accessory cavities, often bear the typical character of neuralgia. In affections of the frontal sinus, supra-orbital neuralgia is a constant symptom. Lightning darting pains radiating into the teeth of the upper jaw, the skin of the cheeks, and oral mucous membrane, are almost constantly observed after operative measures, or in consequence of neuralgia of the second branch of the trigeminus.

Treatment.—Treatment in the first place depends on the cause.

In cases of anæsthesia which are not central in origin, galvanism of the mucous membrane should be tried. For hyperæsthesia and neuralgia, the prevention of irritation is recommended, as well as destruction of the diseased parts of the mucous membrane by means of the galvanic cautery; while for fits of sneezing, bromide of ethyl and inhalations of chloroform, and inwardly bromide of potassium and insufflations of morphia (gr ⅙) may be tried.

Anosmia, or loss of smell, is the most common disease of the olfactory nerve that we meet with.

It is either total or partial, unilateral or bilateral.

Among the central causes are : congenital absence of the olfactory bulbs, ruptures of the bulbs in consequence of blows on the head and concussion of the brain, compression of the brain (Quincke), exudation into the anterior cranial fossa, cerebral abscesses and tumours, embolism of the left median cerebral artery, and atrophy of the olfactory bulbs, and its central origin in later life. Paralysis of the trigeminus of long standing leads to weakening of the power of smell, in consequence of deficient nutrition of the nasal mucous membrane. Paralysis of the facial may also give rise to disturbances of smell, by preventing sniffing, the result of paralysis of the dilator and contractor of the nostrils, and also by paralysis of the orbicularis oculi, in which the conjunctival secretion flows over on to the cheeks, and causes the nasal mucous membrane to become dried up. The commonest of peripheral causes is the prevention of odoriferous materials from influencing the olfactory region. The odoriferous substances are either altogether prevented from entering the olfactory region by neoplasms and foreign bodies, or owing to swelling, hypertrophy, or polypoid degeneration of the mucous membrane, they do not reach the end organs of the olfactory nerve. Further causes of anosmia are: dryness of the Schneiderian membrane, absence of pigment in the olfactory cells of Schultze, careless use of the nasal douche, particularly when employing alum, zinc, or carbolic lotions, and inhalation of ammonia vapour and ether. Against the theory, that in all probability, from too prolonged or intense a use of the sense of smell, diminution in its power takes place from over-stimulation of the olfactory nerves, there stands the fact that in the case of flayers, anatomists, dissecting-room porters, and latrine cleaners, no such

s

impairment occurs. Intermittent anosmia was observed by Raynaud in the case of a lady, who, every day at four o'clock in the afternoon, lost her power of smell for eighteen hours.

The prognosis of anosmia is rather unfavourable. It is most favourable in acute and chronic catarrh, nasal polypi, foreign bodies, &c. Morell Mackenzie says he has never seen the power return, when the affection has lasted more than two years.

The diagnosis is made by testing the sense of smell. First one and then the other nostril is closed, whilst odoriferous substances such as peppermint oil, cinnamon drops, valerian, oil of turpentine, carbolic acid, creasote, eau de Cologne, &c., are held to the nostril to be examined.

Treatment.—The treatment must be guided by the cause. Anosmia from injuries to the head, hæmorrhages, exudations, and emboli, often improves in time without any treatment. Catarrh and neoplasms must be treated according to the prescribed methods. In true functional anosmia, a trial of the constant current is indicated; unfortunately a very strong current is required to excite the olfactory fibres (Althaus), so much so that violent pains, and swelling of the mucous membrane, are unavoidable. Subcutaneous injections of strychnine, painting the mucous membrane with strychnine, gr. 5, to olive oil ʒi, or insufflations of strychnine, gr. $\frac{1}{24}-\frac{1}{12}$, to starch, grs. 3, twice a day are sometimes beneficial. In intermittent anosmia, quinine, salicylate of soda, tincture of eucalyptus, arsenic, and iron are recommended.

Exaggeration of the sense of smell, hyperosmia or hyperæsthesia olfactoria is sometimes, like sharpening of the sense of taste, a physiological symptom. Wild uncivilised people, according to the reports of different travellers, appear to possess an extraordinarily fine sense of smell. The case of the boy James Mitchell is well known, who, born without any other sense, cultivated his sense of smell so widely, that he was able by means of it to distinguish persons and things, just as a dog. As a pathological symptom, hyperosmia occurs pretty often in hysterical, anæmic persons, and particularly often in pregnant women. They smell the least quantity of odorous material, which, under ordinary circumstances, could not be perceived by them.

Perverted sense of smell, parosmia, paræsthesia olfactoria, or kakosmia subjectiva, is also comparatively often met with.

The causes are pretty much the same as those of anosmia, such as acute and chronic catarrh, polypi, &c., tumours of the brain, syphilis of the brain, apoplectic attacks, cerebral hyperæmia, atheroma of the cerebral vessels, and inflammatory affections of the olfactory nerves. With some epileptics perverted sense of smell precedes the attack as an aura, and very often in insane patients there are illusions and hallucinations of smell. In this connection also, are the aberrations of smell met with in hysterical and pregnant women, who exhibit a predilection to the smell of burnt feathers, valerian, &c. In cases of parosmia, either smells are perceived, which do not really exist, or what occurs more frequently, existing smells appear altered, being generally perceived as of an unpleasant, burnt, or putrid character (*allotrioosmia*). I have repeatedly observed kakosmia subjectiva in hypochondriacs, among others a young intellectual, and perfectly healthy workman, who believed that he emitted such an offensive smell from his anus, that his comrades avoided him, while an older nervous official could not get rid of the idea that he diffused a pestilential smell from his nose.

The prognosis of parosmia is, to say the least, doubtful, but is most favourable when there are material alterations of the nasal mucous membrane, and in hysterical and pregnant women.

Treatment.—Treatment here also depends on the etiology. Insufflations of nitrate of silver, the galvanic cautery in acute and chronic catarrh, operations upon neoplasms, and treatment of epilepsy and hysteria, soonest lead to a successful issue. With the above mentioned hypochondriacs, I tried aromatic snuffs, touching the olfactory region with the galvano-caustic point, and inwardly bromide of potash, but without result.

DISEASES OF THE ACCESSORY CAVITIES.

Of the adjacent or accessory cavities, *the maxillary sinus* is most frequently diseased.

The causes are to be found in the great frequency of pathological changes in the teeth and their alveoli, and also in the nose itself. The diseases which usually give rise to affections of the sinus are, alveolar periostitis, dental caries, defects of teething, ingrowth of teeth into the sinus, foreign bodies, injuries, and also primary and secondary inflammation of the nose.

Acute and chronic catarrh spreads through the aperture of communication to the mucous membrane of the antrum. The lumen of the opening may remain permeable, be narrowed, or completely closed, becoming more easily closed when oval or slit-like, than when round and wide. At first, according to Zuckerkandl, the lining membrane of the sinus is seen to be injected in its whole thickness, reddened, and sometimes ecchymosed. As yet there is no secretion, but later on, when the mucous membrane is considerably swollen, exudation appears. The mucous membrane is then seen to be œdematous and gelatinous, and its surface covered with large dropsical swellings, the sinus also contains free fluid, partly purulent, partly mucous, or hæmorrhagic in character, and of hyaline consistence. If the sinus be filled with a great mass of mucus, then one speaks of hydrops antri Highmori, although, from an anatomical point of view, this designation is not considered a happy one.

The symptoms of acute or chronic inflammation of the antrum, when effusion is scanty, are very various. Some patients complain of dull pain, deep in the upper jaw, which is increased by pressure on the bones, or by mental effort. Others complain of toothache, sometimes continual, sometimes intermittent, chiefly in the back teeth of the upper jaw. This is explained on the one hand by the inflammation, on the other by the pressure which the collected fluid exerts on the dental nerve; in consequence of this pressure the teeth may atrophy, become carious and fall out, or the patients, wearied by the constantly returning pains, get them extracted. If the aperture of communication is not closed up, then, with the body in a proper position, the fluid may flow out, and the pain disappear for a time. If the cavity contain air and fluid, on shaking the head, sounds may be heard as in pneumothorax. When the sinus is completely filled with fluid, and the opening entirely closed, the exudation sometimes exerts so much pressure on the walls of the cavity, that it becomes dilated in various directions.

If the accumulated fluid is pus, the affection is called empyema of the antrum, which, when the aperture into the nose is completely closed, is generally accompanied by the most violent pains, with intermittent fever and rigors, general discomfort, œdema of the cheeks, and the feeling of lengthening of the teeth. Very often prominence and enlargement of the sinus is caused. The canine fossa appears pushed forward, is painful on pressure, and becomes distinctly fluctuating on the bones becoming thin. The anterior part of the hard palate, immediately over the teeth, may also appear prominent, causing the patients to complain of discomfort and pain on chewing. Sometimes the entire half of the hard palate becomes uniformly arched forward, and is sensitive to touch, so that chewing becomes still more difficult, and swallowing and speaking in the highest degree painful. As the hard palate forms the floor of the nasal cavity, gradual, often complete, closure of the nose may occur.

The difference in thickness of the walls of the sinus, causes a difference in the manifestation of the objective symptoms.

If the walls of the canine fossa, and of the hard palate, be very thick, the collected fluid presses in that direction in which there is least resistance. Hence the prominence sometimes appears on the zygoma under the eye. The orbit is forced out of its normal position, exophthalmos arises, and atrophy of the optic nerve when the pressure becomes stronger. If the patients do not seek help in good time, the pus perforates through, and fistulous openings are left after abatement of the pain. If the aperture of communication with the nose be free, there is often no pain, and a great quantity of thick, exceedingly fœtid pus may be emptied from the sinus, by placing the body in a suitable position,—generally by bending the head forwards, and to one side.

Phlegmon of the antrum of Highmore, according to Weichselbaum, appears to be a very rare disease.

It is characterised as an inflammation of the lining membrane of the sinus, with deposition of disseminated or confluent and putrefying fibrinous patches in the membrane itself. The antrum seems to be most often and most seriously attacked, though the other accessory cavities also suffer. The nose itself is either unaffected, or shows only insignificant changes.

Both in the primary and secondary development of the affection, especially in phlegmonous processes of the tongue, or of the throat, or in croup of the larynx and pharynx, the symptoms are excessively severe, and consist of disturbances of the function of the brain, and also in fever and pains in the neighbourhood of the maxillary and frontal sinuses, which may cause death in from two to four days. As complications of the primary form, Weichselbaum found swelling of the spleen, cloudy swelling, with partial fatty degeneration of the liver and kidneys, and ecchymoses of the pleura and pericardium.

Facial erysipelas also sometimes leads to phlegmonous inflammation of the lining of the sinus, which, however, is distinguished from the above form by the absence of fibrinous deposits. Vice versa, phlegmon of the sinuses leads to facial erysipelas, this being especially the case in erysipelas accompanying typhoid fever.

Diphtheria of the maxillary sinus, according to Zuckerkandl, may run its course without membrane formation, as the so-called mucous membrane of the sinus, in virtue of its anatomical structure, is incapable of forming a pseudo-membrane. Ecchymoses and œdema only take place.

The maxillary sinus is not infrequently the seat of neoplasms.

Mucous polypi are the most common, and are divided by Zuckerkandl into pedunculated tumours, tumours stretched bridge-like between the walls, and flat mucous membrane tumours. Cysts arise from the mucous glands, and vary much in size. Besides these, fibromata, osteomata, osteophytes, ossifying fibromata, myxochondro-sarcomata, and epitheliomata have been observed. They excite very much the same symptoms as exudation.

The diseases of *the frontal sinus* are pretty much the same as those of the antrum.

The frontal sinuses participate much oftener in acute and chronic nasal catarrh, and sometimes also in phlegmonous angina. Inflammation of the frontal sinus causes severe frontal headache, which sometimes precedes the outbreak of nasal catarrh, sometimes accompanies or follows it. In some persons the frontal sinus suffers with every catarrh. The pain, which the patient at one time localises near the glabella, at another deep down in the eye or the eyebrows, is sometimes continuous, sometimes intermittent, and of a distinctly neuralgic character. It is increased by mental effort, movement of the body, the use of alcohol, or by pressure on the bones, and may even lead to nausea and vomiting, so that the disease very strongly resembles megrim. Supra-orbital neuralgia is hardly ever absent in inflammation of the sinus. Hartmann considers this due to the resorption of a part of the air shut up in the frontal sinus, thus giving rise to diminished atmospheric pressure, increased dilatation of the vessels, swelling of the mucous membrane, and increased secretion. In spite of the frequency of this symptom in inflammation of the frontal sinus, it must not be forgotten that it may originate reflexly from the nasal mucous membrane.

If, at this stage, the sinus membrane be examined, it appears considerably swollen, œdematous, and ecchymosed. An exudation of muco-purulent fluid arises generally only after a considerable time. The aperture of communication may remain patent, or be partially closed. In the latter case a large quantity of serous or purulent fluid may be suddenly evacuated, and cause instantaneous relief, as the cases related by Fischer and Paget prove.

When the excretory duct is entirely and permanently closed, there arises, if the secretion be mucous, hydrops, and if purulent, empyema of the frontal sinus. In both conditions, there is in consequence an exaggeration of the subjective symptoms, with fever, disinclination to work, general discomfort, stiffness in the neck, giddiness, and double vision. Later on the walls of the sinus become bulged out, which, o course, will also occur if the sinus be filled with blood, as Steiner once observed.

Severe supra-orbital neuralgia is usually present, which diminishes or disappears, as Hartmann pointed out, when the accumulated secretions become evacuated.

Dilatation of the frontal sinus becomes apparent, by arching forward of the inner and upper wall of the orbit. The eye is pushed downwards and outwards, disturbed vision even to amaurosis arises; the upper eyelid and the skin over the eyebrows look œdematous and inflamed. When the dilatation of the sinus is very considerable, the pressure on the frontal lobe may cause unilateral paralysis, as the case described by Otto shows. The fact that cerebral symptoms are sometimes absent, is considered by Richter to be due to the brain accommodating itself to the slow and gradually increasing pressure of the dilating sinus. Sometimes the abscess perforates near the inner angle of the eye, sometimes it bursts into the orbit, and forms there, as in Buller's case, an elastic fluctuating swelling, and in very unfavourable cases breaks through into the brain.

The conditions which lead to empyema of the sinus, may also arise from external influences. Apart from fractures and injuries to the sinus from foreign bodies, other affections, and especially syphilis, may cause extensive destruction of the frontal bone and opening into the sinus, after which, partial and general emphysema of the skin has repeatedly been observed.

Herniæ of the frontal sinus are very rare occurrences. One case is recorded by Rizet, another by König. In the latter, an ulcer probably of syphilitic origin had healed; formerly on blowing the nose it had been observed that air passed through, but now on forced expiration a subcutaneous swelling appeared, soft and reducible, which could be kept down by a truss-like arrangement. In Rizet's case syphilis was also the cause; the hernia disappeared after removal of a necrosed piece of bone.

Neoplasms also occur in the frontal sinus. Mucous polypi and cysts are certainly found, but osteomata chiefly occur, arising, as Bornhaupt, Arnold, Banga, and Björken have shown, from the inner surface of the sinus, and consisting partly of compact, partly of spongy bone tissue; they usually remain local, but may penetrate the skull. The symptoms are caused by the size, situation, and growth of the tumours.

We turn now to diseases of the *sphenoidal cells or sinuses*.

Retention of secretion takes place very easily on account of the unavoidable position of the excretory aperture.

Chronic purulent inflammation of the sphenoidal and ethmoidal sinuses is considered by Michel to be the cause of ozæna. Although it cannot be denied that the sphenoidal sinus is occasionally affected in ozæna, and a fœtid secretion discharged, yet it has certainly been proved that genuine ozæna depends on chronic catarrh of the nose accompanied by atrophy of the mucous membrane.

Mucous and purulent accumulations have been observed in the sphenoidal sinus, in tubercular and cerebro-spinal meningitis.

Tuberculosis of the sinuses does not seem to occur, according to Weichselbaum, even in cases of pronounced tuberculosis of the nose. I have found only a single contradictory note by Tornwaldt, who relates that Professor Neumann had once the opportunity of examining fungoid granulations from the antrum in a case of caries in the upper jaw, and that these were permeated with tubercles.

The symptoms which empyema of the sphenoidal sinus excite, may be exactly the same as those of the other cavities. Among others, Rouge relates the case of a lady, who suffered from snuffling, pains in the upper teeth, exophthalmos, strabismus, and later from left-sided deafness and blindness. Thinking the antrum was the seat of the trouble he resected it, but found it quite healthy. The sphenoidal sinus, however, was full of pus.

Neoplasms, such as cysts, mucous and osseous tumours, seldom come under observation.

Diseases of the *ethmoidal sinuses* occur sometimes alone, sometimes in combinations with affections of the other sinuses.

Acute and chronic catarrh of the nose very often gives rise to inflammation of these cavities. Expansion of the ethmoid cells leads to protrusion of the orbit outwards and downwards, to exophthalmos, pressure on the brain, and obstruction of the nose on the affected side. Fractures of the ethmoid, with opening of individual cells, or caries, result in emphysema of the orbits and eyelids. Zuckerkandl considers it probable, that congenital defects and dehiscences of the orbital roof or holes in the bones, cause orbital emphysema on violently blowing the nose, or on blowing with the nose held shut.

Neoplasms, especially mucous polypi of the nose, often have their origin in the ethmoid cells. Ivory exostoses and other tumours also come under observation. To be quite complete, it may be mentioned, that in all these accessory cavities, Virchow has observed a chalky degeneration of the mucous membrane, which generally proceeds to the nose. They are, too, sometimes the seat of parasites, thread worms, and maggots.

The diagnosis of diseases of the accessory cavities depends on the subjective and objective symptoms. The attempts of W. Zenker and Czernicki, to bring auscultation and percussion to help the diagnosis, must be considered out of the question, even although they are not altogether without result in cases of extreme attenuation of the bones, which are so rarely met with, as hardly to deserve consideration. How easy it is to mistake the localisation of the affection, the above-mentioned case of Rouge's shows, where all the symptoms pointed to the maxillary sinus. The diagnosis is the more certain, the more distinctly the objective changes,—tenderness, prominence of the sinus wall, protrusion of neighbouring organs, and sudden gush of a great quantity of fœtid pus,—are developed. It is very difficult, sometimes even impossible, to distinguish whether the subjective symptoms, especially neuralgia of the trigeminus, are of a reflex nature, or caused by material changes of the sinus walls.

Treatment.—Although the treatment of these cavities is still very defective, yet it cannot be said to be altogether without prospect.

First of all, the etiological factors must be considered, all carious teeth and foreign bodies therefore removed, and other anomalies affecting the teeth and alveoli attended to. As chronic swellings of the nasal mucous membrane, and especially of the middle turbinated bone, or polypoid excrescences or mucous polypi, may be the causes of very many diseases of the sinuses, their removal must be accomplished according to the prescribed rules.

The attempt to empty the sinuses by bending the head forwards or to one side, succeeds, as a rule, only when the excretory ducts are quite patent. In order to open them when they are narrowed or closed up, nasal douching is recommended, more especially syringing by means of a ball syringe Michel recommends that after the first syringeful of water has flowed in, the nose be shut with the thumb and first finger, the head bent forwards and then held downwards for a few minutes.

Hartmann uses compressed air in the form of Politzer's air-douche for opening the closed sinus apertures; if the symptoms are temporarily relieved thereby, they are most probably caused by some coexisting nasal affection, but when the results are negative, this seems questionable. Ziem rejects the nasal douche, because putrefactive particles from the nose may get into the ear and there excite deleterious influences. He therefore introduces behind the palate a bent catheter, which is connected to a foot-bellows by an indiarubber tube, and thus causes condensation of the air in the nose and sinuses from the nasal fossæ, the patient at the same time shutting one or both nostrils with the finger; by this means a condensation of air is obtained sufficient to expel the secretion without danger to the ear.

A further proposal has been made, to gain egress for the collected secretion by introducing probes and tubes into the aperture of communication of the sinuses, and to bring remedies into direct contact with the diseased mucous membrane. Although the practicability of this manipulation on the living subject cannot be doubted, my experience is, that it is extremely difficult to perform, and often impossible. The draining of these cavities, previously recommended by Kessel, was lately carried into practice by Jurasz, Hartmann, and others. For the purpose a tube is required 10 to 30 centimetres long, 2 to 2½ millimetres thick, slightly bent forwards, connected by means of an india-rubber tube to a balloon, out of which the fluid to be injected,—generally carbolic lotion,—is driven. Hartmann was able to cure several patients by washing out the sphenoidal and maxillary sinuses, after all other methods had proved useless. Michel considered removal of the middle spongy bone necessary for carrying out the drainage; Hartmann also in one case had to resect the anterior end of the middle spongy bone by means of a cutting ring forceps, before he could gain access for his canula.

If by these methods the accumulated secretion in the pneumatic spaces cannot be removed, then more formidable operations are necessary.

Opening the maxillary sinus can be best accomplished from the alveoli of the molar teeth, because the opening thus made reaches the deepest part of the sinus and renders continuous flow of the secretion possible without stagnation. The first or second molar tooth, if necesssary both, are extracted, and through their alveoli a three-edged stilette, or the point of a pair of scissors, is carefully introduced, with a rotatory movement from below upwards, towards the inner angle of the eye; that the sinus has been reached will be known by the free movement of the stilette and outflow of the secretion. The latter being removed, the sinus is washed out with some disinfectant solution. To restore the mucous membrane to its normal condition, after cleansing with lotions containing boracic or carbolic acid, chlorate or permanganate of potash, it is recommended to bring the walls into contact with an alcoholic solution of iodine. Care must be taken about the concentration, for, as I once saw, acute iodism may be set up by injecting even a very dilute solution. If the opening remain too long, then a drainage tube must be inserted, or a silver canula with a stopcock, which can be fastened by a caoutchouc plate to the palate or to the neighbouring teeth.

Opening the frontal sinus requires incising the skin over the eyebrow or towards the bridge of the nose, and pushing aside the periosteum. A piece of bone is next removed by means of a chisel or a trepan, and the duct into the nose dilated with probes. If the duct cannot be found, a trocar is pushed from the opened up cavity into the nose, and a drainage tube inserted.

Opening the ethmoid sinus is accomplished either from the outside by splitting the nose, or from the inside by pushing a trocar or a strong injection needle through, from behind and above, between the middle and superior spongy bones.

It is hardly justifiable to open the sphenoidal sinus because of the great danger of injuring the brain. The shortest way to the cavity would be from the naso-pharynx, by opening the fornix immediately over the superior border of the nasal fossæ. An opening into this sinus, made by nature, was once observed by Stoerk in a scrofulous boy.

APPENDIX.

Page 34. ZITTMANN'S DECOCTION.

The German Pharmacopœia describes two preparations bearing the name, Decoctum Sarsæ Co., viz., *fortius* and *mitius*. The former consists of sarsaparilla, sugar, alum, aniseed, fennel, senna and liquorice root. Zittmann's decoction is the above boiled with calomel and red sulphuret of mercury tied up in a rag, but without sugar and alum. *Mitius* is made with the residue of the fortius, and contains sarsaparilla, lemon peel, cinnamon, cardamom seeds, and liquorice root.

Page 34. IODIDE OF POTASSIUM IN MALIGNANT DISEASE.

It is by no means an absolute rule that iodide of potassium does not influence carcinoma in the least degree. In some cases of malignant disease, pain is considerably diminished by large doses of the iodide, probably in consequence of diminished blood pressure, as well as from the general sedative action of the drug.

Page 42. THRUSH IN CHILDREN.

Vogel, writing on this subject in "Ziemssen's Encyclopædia" (vol. vi. p. 807), strongly recommends that a healthy wet nurse be got for children affected with thrush in the mouth, and says that, in his experience, this treatment has generally been followed by wonderfully good results.

Page 51. LESIONS OF CHORDA TYMPANI.

Urbantschitsch, in his "Lehrbuch der Ohrenheilkunde" (p. 413), points out that lesions of this nerve affecting taste are very apt to occur in the course of chronic suppuration of the middle ear. He found impaired taste in a considerable number of persons suffering from that affection. Injuries to the nerve from operation and other causes also bring about a similar result, viz., loss of taste in the anterior portion of the tongue on the affected side.

Page 62. EAR AFFECTIONS AFTER MUMPS.

Well marked and even complete deafness, as a result of parotitis, is looked upon in this country as a pretty common occurrence. In Germany and America it is still considered somewhat rare. Toynbee thought it was probably due to an affection of the labyrinth, the exact nature of which still remains unknown, and the observations of other aurists seem to confirm his theory. Roosa found middle ear disease in several cases, though not enough to account for the degree of deafness present. Sometimes both ears are affected, but generally only one.

Page 63. BRUN'S COTTON WOOL.

This preparation is made in Geneva, where it is known as "Ouate anti-rheumatisme." It is impregnated with pine oil, and resembles the wool known in this country as "Lairitz's pine wool wadding." It was first introduced into practice by the late Professor Bruns of Tübingen, and is much used in Germany as an antiseptic dressing.

Page 85. VON TROELTSCH'S METHOD OF GARGLING.

The head being bent backwards, the fluid is allowed to flow into the back of the throat, when repeated forcible movements of swallowing are to be made without actually swallowing the fluid, which is forced up again at the last moment. In this way contraction of the pharynx takes place and powerful displacement of superficial parts; mucus is forced out from the glands, and any adherent viscid secretion is rubbed off. Von Troeltsch very justly extols the remedial gymnastic significance of systematic practice of this kind in insufficiency of the muscles of the Eustachian tube (levator and tensor palati) in cases of hypertrophy of the mucous membrane. The parts reached by this method are the posterior surface of the soft palate, the walls of the lower pharynx, the floor of the orifices of the Eustachian tubes, and the tubes themselves (Wendt in "Ziemssen's Encyclopædia," vol. vii. p. 27).

Page 106. ADENOID VEGETATIONS.

Tinnitus is rarely, if ever, complained of by children affected with deafness from the presence of adenoid vegetations, in fact, is hardly ever complained of by children at all.

Page 170. LIQ. AMMON. ANIS.

The Liquor Ammonii Anisatus of the German Pharmacopœia is made as follows :—Oil of anise, 1 part, is dissolved in rectified spirits, 24 parts, and then mixed with strong liquor ammonii, 5 parts. Dose, 5–15 mins.

Page 201. NASAL CARTILAGES.

The "triangular cartilages" are better known in this country as the upper lateral cartilages. They go to form the tip and alæ of the nose. The cartilages of the alæ nasi, or lower lateral cartilages, curve inwards upon themselves, touch each other in the middle line at the tip, and go to form the anterior and lateral boundaries of the orifice of the nostrils.

Page 213. NASAL SPRAYS.

M. Mackenzie's anterior nasal spray is one of the best of its kind. It consists of a silver pipe three or four inches long, ending in a fine perforated point and provided with a piece of tubing and a hand ball. The pipe passes through a cork which fits an ordinary bottle. From its length the pipe can be passed for some distance into the nares, which is an advantage. The same tube and hand-ball can be adapted to the posterior nasal and laryngeal sprays, which only differ from the other in having appropriate curves.

Page 258. HÆMORRHAGE.

Dr John Duncan, of Edinburgh, has lately introduced a new and safe method of transfusion, which he recommends in cases of primary amputation, where life is in danger from loss of blood, and also in all major amputations. The blood lost by the patient is as far as possible collected in a vessel containing a quantity of saline solution (5 per cent. solution of phosphate of soda, in the proportion of

1 part to 3 of blood)—to prevent coagulation—and, having been brought to the temperature of the body, is reinjected into a vein. The blood, and all instruments, &c., used for the operation, must be carefully rendered aseptic. This method might prove useful in severe cases of recurring epistaxis; the blood being reinjected into one of the superficial veins of the arm.

Lately it has been suggested by some writers, that in cases of spontaneous hæmorrhage, a rhinoscopic examination be made to discover the spot from which the bleeding arises, and when found, that astringents or caustics, such as perchloride of iron, nitrate of silver, or if necessary, the galvanic cautery, be applied to it. By doing this, these writers have been able permanently to cure several cases of recurring nasal hæmorrhage, which up to that time had resisted all the usual methods.

Page 268. Nasal Polypi.

In the case of polypi with thick pedicles, Jarvis' (of New York) nasal ecraseur will be found useful. It is a straight metal tube, from 6 to 8 inches long, smooth at the distal, and grooved at the proximal end. Over the grooved part a second canula is fitted, and along the screw runs a small wheel which, when rotated, pushes the outer canula before it. The wire runs through the entire length of the inner canula, and is fixed to two small pins at the end of the outer tube. The loop can be adjusted to any size required. The instrument is worked either by slowly rotating the wheel or by quickly drawing back the outer tube. Many modifications of this instrument have been made. M'Bride's (of Edinburgh) "ecraseur snare" also answers well in difficult cases, such as Dr Schech describes. It consists of a snare, the wire of which is attached to a bar which slides upon a screw. The instrument may be used as an ordinary snare, but if the growth be found too resistant, the wire can be drawn home by a nut which is screwed up against the bar. This instrument is not so delicate as Jarvis', is more powerful, and also cheaper.

In cases where difficulty is experienced in the removal of nasal polypi situated far back, by their slipping into the naso-pharynx, Berthold finds that by simply plugging the posterior nares, the difficulty is overcome, and the operation rendered much easier.

Dr B. W. Richardson applies cotton wadding saturated with sodium ethylate to the pedicles of polypi for several minutes, by means of a forceps, and states that, on withdrawing the plug, the polypi are ejected by blowing the nose. The stumps are then touched with the same remedy, which for a short time causes burning pain. After using this method he has never seen recurrence, inflammation, or hæmorrhage take place.

COCAINE.

A work of this kind can hardly be considered complete without some notice being given to Cocaine, the active principle of the leaves of the Erythroxylon Coca, which has so lately come into general use as a local anæsthetic. It is usually employed as the hydrochlorate of cocaine, a salt very soluble in water, solutions of which, varying from 2 to 40 per cent., may be used. For ordinary purposes a 4 per cent. solution is generally sufficient, and may be applied by means of a brush, spray, or sponge, or in the form of a powder, or of cotton wool pellets impregnated with it. To it should be added from $\frac{1}{2}$ to 1 per cent. boracic acid, to preserve the alkaloid from a microscopic plant which is said to grow in it and spoil it.

When applied to mucous membrane, cocaine produces anæsthesia of the part for twenty minutes or more, the capillaries contract, causing the part to appear pale, and there is also dryness and suspension of functional activity. All persons are not equally susceptible to the influence of the drug, as is also the case with other narcotics, and in some cases symptoms of poisoning have occurred after using a solution which will produce little or no effect in the case of persons less easily influenced.

Among the diseases in which its use is followed by benefit are tuberculosis and carcinoma. A 4 per cent. solution painted over the parts will generally diminish the dysphagia so often present. The uvula and small growths in the pharynx have been painlessly removed after its use. In a case of catarrhal inflammation of the pharynx the application of a 10 per cent. solution gave great relief. In phlegmon of the pharynx its use has proved beneficial. In acute nasal catarrh, the irritation and congestion of the mucous membrane are dispelled, in some cases, in a few hours. Three applications, at intervals of half an hour or so, have been known to avert an attack. The astringent action of the drug will produce absorption of the hypertrophied mucous membrane in chronic nasal catarrh.

In that condition of the mucous membrane covering the inferior turbinated bodies known as "erection," the application of cocaine in some cases immediately causes a return to the normal. In "rose catarrh" or "rose fever," and in "nervous coryza," associated with paroxysms of sneezing, the drug has been employed with very satisfactory results. In the case of reflex neuroses, due to nasal affections, cocaine may be used to confirm the diagnosis, since the parts of the nasal passages to which it is applied will no longer be capable of producing the symptoms, such as fits of sneezing, asthma, &c., which their former condition set up.

It is very useful in operations on the nasal cavity, not only producing anæsthesia, but also diminishing congestion. Bosworth and Jarvis have employed it in removing polypi, and in operating on deviations of the septum, with good results. In some cases of hay fever the application of cocaine to the nasal mucous membrane has been followed by marked benefit; and if dropped into the eyes will at once relieve the intense itching so often met with in that affection. One practitioner finds the following give great relief:—℞ Cocaine, grs. 4; Collodion, flex. ʒi.; Ol. Ricini ʒiv., painted inside the nostrils, several times a day. Another uses it in the form of powder:— ℞ Acidi boracis, ʒss.; sod. salicyl., grs. 40; Cocaine mur., grs. 2. A small quantity to be blown into the nose. This formula is also said to cut short an attack of acute nasal catarrh if used early, *i.e.*, during the sneezing stage.

As a conclusion to these notes on cocaine, I cannot do better than give in full the concise and exhaustive summary of its uses by Dessar of Wurzburg, translated by Semon—

Solutions of cocaine secure,

1. Diminution of tactile sensibility. This is useful (*a*) to facilitate laryngoscopic examination in cases of hyperæsthesia by abolition of reflex phenomena; (*b*) in posterior rhinoscopy; (*c*) to abolish the augmented sensibility in cases of swelling of the nasal mucous membrane.

2. Diminution of painful sensations (*a*) in operations, (*b*) in divers

examinations, executed in any part of the larynx, pharynx, and nose.

3. Abolition of dysphagia in cases of stenosis, produced by tumours, of phthisis and syphilis (pharyngeal and laryngeal), perichondritis (laryngeal), tonsillitis.

4. Ischæmia of much injected mucous membranes.

5. Diminution of profuse hæmorrhages.

6. A certain diagnosis in cases of nasal reflex neuroses (asthma, different forms of neuralgia, hay fever, epilepsy).

Dr Semon, giving his own results, has found it most useful (1) in tonsillitis, in which it often abolishes, *for a time*, as if by magic, the dysphagia, and enables the poor patients, who have not been able to swallow even fluids for days, to comfortably enjoy a good draught of milk or beef tea; (2) in tonsillotomy, which operation is really rendered perfectly painless by the previous twice repeated application of a 20 per cent. solution to the tonsils and their whole neighbourhood; (3) in uvulatomy; (4) in removal of laryngeal growths; (5) in cauterising the nasal mucous membrane with the galvano-cautery; (6) in diminishing pharyngeal hyper-irritability for purposes of laryngeal and rhinoscopic examination; (7) in acute coryza of adults and infants; (8) in laryngeal pthisis In hay fever Dr Semon has not had any satisfactory results, the effect being very temporary. In cases of nasal polypus his results have varied very much, some patients having no pain, and others as much as when it was not applied.

FERRIER'S SNUFF.

Another remedy found useful, both in coryza and hay fever, which may be mentioned here, is Ferrier's snuff or Pulv. Bismuthi co. :—R Morph. hydrochlor. grs. ii.; powdered acacia, ʒii.; Bism. Subnitr., ʒvi. From 2 to 4 drachms of this powder may be used as snuff in twenty-four hours.

GLYCERINUM ALUMINIS.

Mr Parker, surgeon to the East London Children's Hospital, strongly recommends the use of this preparation. It is made by dissolving 1 part of alum in 5 of glycerine, by the aid of gentle heat, this being four or five times stronger than a saturated watery solution. Having given it a prolonged trial, Mr Parker finds it to be quite as powerful an astringent as tannin, and far less disagreeable in taste; also that it is quite compatible with iron. It is very efficacious in chronic pharyngeal catarrh, so common in children; diluted with water it forms a useful gargle, injection, or lotion. (*Brit. Med. Jour.*, 1885, Vol. I., p. 178).

TINCTURE OF BENZOIN.

Dr Kebbell (*Brit. Med. Jour.* 1885, Vol. I., p. 430) has for three years employed tincture of benzoin as an inhalation in cases of nasal catarrh and influenza. It is inhaled directly from the bottle through each nostril separately, the other being held shut, and long inspirations should be taken. If used early enough it will cut short an attack, and in the later stages will relieve the stuffy, hot, and uncomfortable sensations in the nose, and cause the nostrils to become cool and clear, and the mucus to assume the character it has at the end of a week, in cases where the affection runs on without treatment. In one case only did it fail to shorten the symptoms, which were unusually severe, still from its use they were very much mitigated.

SYRUP OF HYDROIODIC ACID.

This preparation is recommended by W. Judkins as a remedy for hay fever, giving it in drachm doses every hour or every two hours.

NASAL BOUGIES.

Dr Mackenzie (Edin.) recommends the use of bougies containing Extr. Belladon. in cases where nasal reflex neuroses (asthma, hay fever, sneezing, cough, &c.), are slight. The nose having been cleansed by means of a spray, a gelato-glycerine bougie, containing $\frac{1}{14}-\frac{1}{8}$ gr. extr. belladon. is introduced into it at bedtime, and allowed to dissolve. Dr James Ross gives a new formula for the basis of these bougies:—
℞. Gelatin. prep. ʒi; aq. dest. ʒiss, to be soaked through for 12 hours, when glycerini ʒiss is added, and the mixture dissolved in a water-bath.

INDEX.

INDEX.

	PAGE
Absence of nose,	216
,, tongue, congenital,	7, 206
Accessory cavities to nares, *see* pneumatic spaces.	
Acute nasal catarrh,	220
Adenoid tissue,	72
,, vegetations,	104, 287
Adhesions of nostrils,	216
Aditus pharyngis,	74
Ageustia,	51
Alæ nasi,	201
,, diphtheria of,	278
,, diseases of,	275
,, phlegmon of,	277
Albuminuria after diphtheria,	157
Allotriogeustia,	52
Allotrio-osmia,	275
Anæsthesia dolorosa,	192
,, gustatoria,	51
,, of the mouth,	50
,, of the nasal mucous membrane,	272
,, of the pharynx,	192
Anæsthetics, in removing adenoid vegetations,	109
Anatomy of the mouth,	1
,, of the nose,	201
,, of the pharynx	71
Angina, arthritica,	95
,, cachectica,	101
,, catarrhalis,	94
,, conenneuse,	140
,, diphtheritica,	144
,, follicularis,	98
,, hæmorrhagica,	101
,, herpetica,	140
,, Ludovici,	67
,, nosocomialis phagedænica,	171
,, pultacea,	101
,, scorbutica,	101
,, toxic,	95
,, ulcerosa,	101
Anomalies of mouth,	7
,, of nose,	216

	PAGE
Anomalies of pharynx,	89
Anosmia,	273
Anterior rhinoscopy,	209
Antrum of Highmore,	206
,, hydrops of,	246, 276
Aphonia after diphtheria,	159
Aphtha, Bednar's,	23
,, of mouth,	15
,, of pharynx,	143
Aphthous stomatitis,	13
Arcades of pharynx,	73
Ascaris lumbricoides in nose,	254
Aspergillus nigricans vel fumigatus,	42
Asthma in chronic pharyngitis,	120
,, ,, rhinitis,	231
,, in hay fever,	225
,, in relation to nasal tumours,	265
Atresia from disease of antrum,	277
,, of nares,	216
Bacillus fasciculatus,	182
Ball spray,	86
Bartolini's ducts,	57
Bednar's aphtha,	23
Bellocq's sound,	258
Benzoin, tinct. of, for catarrh,	292
Berthold, removal of polypi,	288
Bismuth, compound powder of,	291
Black tongue,	42
Blennorrhœa, Stoerk's,	90, 238
Bohn, mucous membrane milia,	24
Bosworth, use of cocaine,	290
Bougies, nasal,	235, 292
Brachylia,	7
Bresgen's insufflator,	86
Bronchial diphtheria,	152
Bronchiolitis exudativa of Curschmann,	231
Brun's cotton wool,	286
Bursa pharyngea,	72
Cachectic angina,	101
Carcinoma of nose,	264
,, of parotid,	66

	PAGE
Cardiac disease in diphtheria,	158
Caspary, transient innocent plaques of the tongue,	25
Catarrh,	220
Catarrhal Stomatitis,	8
Cauterising the nares,	215
,, the pharynx,	86
Cavernous spongy tissue on turbinated bones,	205
Cavum oris,	1
Cheilitis glandularis apostematosa,	12
Chorda tympani, lesions of,	51, 285
Chronic pharyngeal catarrh, paræsthesia from,	194
,, ,, symptoms of,	121
Cleft palate,	89
Cocaine,	289
Complications of diphtheria,	157
Compresses,	83
Concretions in mouth,	65
,, nose,	259
,, salivary glands,	67
,, throat,	185
,, tonsils,	187
Constrictors of pharynx,	197
,, ,, paralysis of,	76
Cooper Rose's intra-nasal plug,	258
Coryza, acute,	220
Coryzariæ,	223
Cough in chronic rhinitis,	231
Croupous angina,	144
Cynanche cellularis maligna,	67
,, contagiosa,	144
,, diphtheria,	147
Deafness after mumps,	286
Deformities of mouth	7
,, nose,	216
,, pharynx,	89
Dessar on uses of cocaine,	290
Dessois' glossophyton,	42
Development, effect of hypertrophy of tonsils on,	123
Deviation of nasal septum,	217
Diphtheria,	144
,, anæsthesia from,	192
,, aphonia after,	159
,, complications of,	150
,, croupous,	147
,, danger of, from tonsillotomy,	130
,, effect on eye,	160
,, effect on speech,	149
,, gangrenous,	155
,, genuine primary superficial,	146
,, heart disease in,	158
,, kidney ,,	157
,, mortality of,	162

	PAGE
Diphtheria of larynx,	151
,, of maxillary sinus,	278
,, of nose,	150, 246
,, of skin and sexual organs,	153
,, of trachea and bronchi,	152
,, paralyses after,	159
,, primary nasal, of infants,	246
,, pulmonary disease in,	158
,, real tissue,	146
,, septic,	154
Diphtheritic pharyngitis,	144
,, rhinitis,	246
,, stomatitis,	27
Duncan's (Dr John) method of transfusion,	282
Dysphagia hysterica,	196
Ear affections after mumps,	286
,, effects of nasal catarrh on,	222
,, in nasal diphtheria,	151
Emmerich, discovered diphtheria bacteria,	146
Ephidrosis parotidea,	65
Epiphora in nasal tumours,	264
Epistaxis,	256, 282
,, vicarious,	256
Epithelial plugs,	48
Epithelioma, hæmorrhage in,	43
,, of mouth,	48
,, of nose,	264
,, of pharynx,	188
,, of tongue,	48
Epithelium of mouth,	3, 48
Epstein, pearly epithelial accumulations,	24
Epulis,	48
Erysipelas of the mouth,	15
,, pharyngis,	141
Ethmoidal sinuses,	207
,, ,, diseases of,	281
,, ,, neoplasms of,	281
Eustachian tubes,	72, 81
,, case of foreign body in,	185
,, closure of,	90
,, cushion,	72, 80
Examination of mouth,	6
,, nose,	208
,, pharynx,	76
Excoriations, superficial, of tongue,	24
Exostoses of septum narium,	264
,, spinal column,	190
Exudative pharyngitis,	140
,, stomatitis,	13
Eye, effects of diphtheria on,	160
Ferrier's snuff,	291
Fibrinous pharyngitis,	143

INDEX.

	PAGE
Fœtid pharyngitis,	118
Foot and mouth disease, in mouth,	15
,, ,, in nose,	252
,, ,, in pharynx,	180
Foreign bodies in nose,	259
,, pharynx,	185
,, salivary glands,	67
Fränkel's nasal speculum,	209
Frenulum labii inferioris,	1
,, superioris,	1
Frontal sinus,	207
,, diseases of,	278
,, empyema of,	264, 280
,, herniæ of,	280
,, neoplasms of,	280
Fur on the tongue,	2
Galvanic cautery,	235
,, description of,	87
Galvano-caustic batteries,	88
Galvano-caustic or electric snare, amputation of tonsils with,	130
,, in removing adenoid vegetations,	110
,, for nasal polypi,	268
Gangrene from diphtheria,	155
Gangrenous pharyngitis,	170
,, stomatitis,	27
Gargles, use of,	85
,, value of, in chronic pharyngitis,	125
Gautier, epithelial exfoliation of the tongue,	25
Genuine ozœna,	239
Gingivitis,	9
Glanders of the mouth,	15
,, nose,	251
,, pharynx,	180
Globus hystericus,	194
Glossitis, phlegmonous,	11
,, syphilitica indurativa,	12
Glossophyton,	42
Glycerine of alum,	291
Goitre, excision of,	191
,, retropharyngeal,	190
Gottstein's nasal plug,	243
Granular pharyngitis,	113
Gravitation abscess,	136
Gullet, paralysis of,	160
Gumma of hard palate,	33
,, mouth,	32
,, nose,	247
,, pharynx,	173
,, soft palate,	173
,, tongue,	32
Gutta rosacea,	231
Hack, reflex neuroses,	230

	PAGE
Hack, superficial exfoliation of the tongue.	25
Hæmatoma of the nose.	256
,, of the pharynx,	183
,, retropharyngeal,	184
Hæmorrhage,	287
,, of mouth,	43
,, of nose,	255
,, of pharynx,	183
Hæmorrhagic angina,	101
Hager and Brand's olfactorium,	224
Hard palate, effect of inflammation of antrum on,	277
,, gumma of,	33
Harelip,	7
Hartmann's nasal brush,	242
,, on neuralgia from inflammation of frontal sinus,	279
Hay fever or catarrh,	224
Health resorts, value of, for chronic pharyngitis,	125
Heart, effects of diphtheria on,	158
Hering, use of chromic acid,	215
Herpes labialis,	14
,, of pharynx.	140
Heubner, nature of diphtheria,	146
Hiatus semilunaris,	203
Hydroiodic acid, syrup of,	292
Hyperæsthesia gustatoria,	52
,, of mouth,	51
,, of nose,	272
,, of pharynx,	193
,, olfactoria,	274
Hyperosmia,	274
Ice, method of applying,	83
Ichthyosis linguæ,	44
Idiopathic mucous patches,	44
,, spasm,	53
,, stomacace,	21
,, ulcers of mouth,	24
Ignipuncture in glossitis,	13
Influenza,	220
Infundibulum,	203
Intermittent anosmia,	274
Iodide of potassium in malignant disease,	285
Irrigator for nose,	212
Isthmus faucium,	73
Jarvis' nasal ecraseur.	288
,, use of cocaine,	290
Johnston, case of nasal diphtheria after operation for polypus,	246
Jurasz, nasal speculum,	209
,, operation on septum,	219
Kakosmia subjectiva,	274

INDEX.

	PAGE
Keratosis linguæ,	44
Kidneys, effects of diphtheria on,	157
Kramer's nasal speculum,	209
Kuhn, treatment of adhesions of soft palate,	93
Lachrymation in chronic rhinitis,	230
Laryngeal diphtheria,	151
Larynx, diphtheria of,	151
,, paralysis of,	160
,, stenosis of, from diphtheria,	159
Lateral bands of pharynx,	114
,, ,, cauterisation of,	127
Lepra of pharynx,	178
Leptothrix buccalis,	3, 22, 42
Leucoplakia oris,	44
Lieblinski, treatment of adhesions of soft palate.	93
Lincoln's concealed electrode,	271
Lips, hypertrophy of,	7
,, paralysis of.	54
Liq. ammon. anis.,	287
Listerine,	243
Lockjaw,	53
Lucilia hominivora,	254
Lupus of nose,	251
,, pharynx,	178
Maas, unilateral macroglossia,	12
M'Bride's nasal ecraseur,	288
Mackenzie's compound alkaline wash,	234
,, pigment in diphtheria,	166
,, anterior nasal spray,	287
Macroglossia,	12
Malformations of mouth,	7
,, nose,	216
,, pharynx,	89
Mandl's solution,	35, 100, 126, 193
Massage,	84
Masseter, paralysis of,	54
,, spasm of,	53
Mastication, muscles of,	5
Maxillary sinus,	206, 275
,, ,, diseases of,	276
,, ,, method of opening,	283
,, ,, neoplasms of,	278
Mayer, method of removing adenoid vegetations,	109
Mercurial ptyalism,	59
,, stomatitis,	18
Micrococcus diphtheriæ,	146
Microstomia,	7
Mitchell, James, case of,	274
Monti, nasal diphtheria in children,	246
Mouth, anatomy of,	1
,, aphtha of,	15
,, epithelium of,	3
,, erysipelas of,	15

	PAGE
Mouth, glanders of,	15
,, malformation, anomalies, and deformities of,	7
,, methods of examining,	6
,, mucous cysts of,	47
,, muscles of,	4
,, nerves of,	4
,, scurvy of,	19
,, syphilis of,	31
,, tuberculosis of,	35
,, variola of,	15
Mucous polypi of maxillary sinus,	278
,, ,, nose,	262
Mumps,	60
Mykosis tonsillaris benigna,	181
Myxadenitis labialis,	12
Nasal bones,	201
,, cartilages,	287
,, cavities, formation of,	202
,, douches,	212
,, fibromata,	263
,, fossæ,	80
,, ,, closure of,	90
,, meatus,	80, 203
,, mucous membrane,	204
,, ,, anæsthesia of,	272
,, ,, hyperæsthesia of,	272
,, papillomata,	263
,, polypi,	262
,, sarcomata,	263
,, septum,	80, 201, 210
,, specula,	209
,, sprays,	287
Naso-pharyngeal polypi,	188
,, ,, removal of,	270
Naso-pharynx,	72
,, chronic catarrh of,	103
,, digital examination of,	82
,, examination of,	78
,, phlegmon of,	132
,, variola of,	142
Neoplasms and tumours of mouth,	44
,, ,, nose,	261
,, ,, pharynx,	188
Neoplasms of ethmoidal sinuses,	281
,, frontal sinus,	280
,, maxillary sinus,	278
,, sphenoidal sinuses,	281
Nervous catarrh,	230
Nerves, diseases of, of mouth,	50
,, ,, nose,	272
,, ,, pharynx,	192
Neuralgia of nose,	272
,, pharynx,	193
Noma or water cancer,	28
Nose, anatomy of,	201
,, bath,	212

INDEX. 299

	PAGE
Nose, deformities, anomalies, and malformations of,	216
,, diphtheria of,	246
,, divisions of,	201
,, examination of,	208
,, general therapeutics of,	212
,, hæmorrhage of,	255
,, parasites in,	253
,, scrofula &c., of,	251
,, septum of, *see* septum.	
,, syphilis of,	247
,, tubercle of,	250
,, tumours of,	261
Nuhn-Blandin gland,	50, 57
Œdematous pharyngitis,	131
Œsophagus, paralysis of,	197
Oidium albicans,	38, 181, 253
Olfactory fissure,	203
,, mucous membrane,	204
,, nerve,	204
,, ,, diseases of,	272
,, region,	204
Open aphtha,	16
Ostium maxillare,	203
Otalgia in nasal catarrh,	222
Oxyuris vermicularis in nose,	254
Ozæna,	239
Papayotin,	165
Papillæ of tongue,	2
Papillitis,	9
Paquelin's thermo-cautery, objection to,	127
Paræsthesia of mouth,	52
,, nose,	274
,, pharynx,	194
,, olfactoria,	274
Paralysis after diphtheria,	159
,, of lips,	54
,, of masseter,	54
,, of œsophagus,	179
,, of pharynx,	196
,, of soft palate,	160, 196
,, of tongue,	54
Parasitic diseases of mouth,	38
,, nose,	253
,, pharynx,	181
Parenchymatous inflammations,	11
Parosmia,	274
Parotid gland,	56
,, concretions in,	67
,, cystic tumours of,	66
,, hypertrophy of,	65
Parotis accessoria,	56
Parotitis, ear disease from,	286
,, idiopathic,	60
,, secondary,	63
"Peenash,"	254

	PAGE
Pemphigus of mouth,	14
,, pharynx,	142
Pentastoma tænioides,	254
Peritonsillitis,	134
Pharyngeal cough,	120
,, croup,	143
,, tonsil,	72
,, ,, hypertrophy of,	104
Pharyngitis, acute catarrhal,	94
,, chronic,	101
,, ,, atrophic,	117
,, ,, hypertrophic,	103
,, diphtheritic,	144
,, exudative,	140
,, fibrinous,	143
,, fœtid,	118
,, gangrenous,	170
,, granular,	113
,, lateralis hypertrophica,	114
,, œdematous,	131
,, phlegmonous,	131
,, scrofulous,	179
,, sicca,	103, 117
,, suppurating,	131
Pharyngo-laryngeal cavity,	74
,, chronic catarrh of,	116
Pharyngomykosis benigna,	161
,, vel leptothricia,	43, 181
Pharyngo-nasal cavity—*see* nasopharynx.	
Pharyngo-oral cavity,	73
Pharynx, anatomy of,	71
,, anæsthesia of,	192
,, aphtha of,	143
,, concretions in,	185
,, divisions of,	71
,, erysipelas of,	141
,, foot and mouth disease,	180
,, foreign bodies in,	185
,, glanders of,	178
,, hæmorrhage of,	183
,, herpes of,	140
,, hyperæsthesia of,	193
,, lepra of,	178
,, lupus of,	178
,, malformations, deformities, and anomalies, &c., of,	89
,, neoplasms of,	188
,, neuralgia of,	193
,, paræsthesia of,	194
,, paralysis ,,	196
,, parasitic diseases of,	181
,, pustules in,	142
,, sarcinæ in,	183
,, spasm of muscles of,	196
,, syphilis of,	172
,, therapeutics of,	83

Pharynx, thrush of,	181
,, tuberculosis of,	176
,, tumours of,	188
,, variola of,	142
,, vasomotor neurosis of,	195
Phlegmonous pharyngitis,	131
,, rhinitis,	244
,, stomatitis,	11
Phlyctenular stomatitis,	13
Plaques, innocent,	25
,, pterygoidiennes,	24
,, syphilitic,	32
Plica salpingo-palatina.	72, 80
,, pharyngea,	72, 80
Plugging the nares,	258
Pneumatic spaces,	206
,, diseases of,	275
,, physiological significance of,	208
Pneumonia after diphtheria,	158
Polypi, nasal,	268, 288
,, naso-pharyngeal,	188, 261
Post-nasal catarrh,	96
Post-pharyngeal abscess,	136
Ptyalism, or salivation,	57
,, in chronic rhinitis,	230
Purulent nasal catarrh,	226
Ranula,	49
Raynaud, case of intermittent anosmia,	274
Reaction from cauterising pharyngeal granulations,	128
Recessus pharyngis,	72
Reduplication of nose,	216
Reflex neuralgia,	230
,, neuroses in chronic nasal catarrh,	230
,, causes of,	232
Retropharyngeal abscess,	136
,, goitre,	190
,, hæmatoma,	184
,, sarcomata,	188
Rhagades,	32
Rhineurynter, or rhinobyon, of St Ange,	258
Rhinitis, acute catarrhal,	220
,, chronic,	227
,, ,, atrophic,	227, 238
,, ,, hypertrophic,	228
,, chronica atrophicans fœtida,	239
,, diphtheritic,	246
,, phlegmonous,	244
Rhinoliths,	260
Rhinoscleroma,	236
Rhinoscopy, anterior,	81, 208
,, posterior,	78, 208
Richardson on sodium ethylate,	289

Rima oris,	1
Rivini, ducts of,	57
Rose fever,	224
Rosenmuller's fossa,	72, 81
Saliva, function of,	57
Salivary calculi,	67
,, glands, diseases of,	56
,, ,, tumours of,	66
Salivation,	57
,, in chronic rhinitis,	230
Sarcinæ in mouth,	43
,, pharynx,	183
Sarcoma of nose,	263
,, tonsils,	188
,, retropharyngeal,	188
Schech's post-nasal forceps,	110
Schneiderian membrane,	204
,, ,, dryness of,	273
Schroetter's dilating tubes,	94
,, use in diphtheria,	168
Schuller, case of new born infant with nasal diphtheria,	246
Scotoma, scintillating, in chronic rhinitis,	230
Scrofula of nose,	251
,, pharynx,	178
Semon, examination of naso-pharynx,	107
,, on use of cocaine,	291
Septic diphtheria,	147, 154
Septum narium,	80, 201, 210
,, crests or spines on,	218, 264
,, deviations of	217
,, phlegmon of,	245
Sexual organs, diphtheria of,	153
Skin, diphtheria of,	153
Smell, nerves of,	204
Snare, for nasal polypi,	266, 288
Sodium ethylate, for polypi,	289
Soft palate, adhesions of,	90
,, chronic catarrh of,	115
,, gumma of,	173
,, paralysis of,	160, 196
,, phlegmon of,	133
,, tumours of,	188
Soor, or thrush of mouth,	38
,, ,, of nose,	253
,, ,, of pharynx,	181
Spasm of constrictors of pharynx,	196
,, masseter,	53
,, œsophagus,	196
,, soft palate,	196
,, tongue,	53
Specula, nasal,	210
,, Zaufal's post-nasal,	81
Sphenoidal sinus,	207
,, diseases of,	280

	PAGE
Sphenoidal sinus, empyema of,	281
,, neoplasms of,	281
,, tuberculosis of,	280
Spinal column, exostoses of,	190
Sprays,	85
Steele, operation on septum,	219
Stenosis of pharynx from cicatrices,	91
Stenson's duct,	56
Stoerk's blennorrhœa of respiratory mucous membrane,	232
,, prognosis of,	242
,, treatment of,	244
Stomacace,	21
Stomatitis, catarrhal,	8
,, diphtheritic,	27
,, exudative,	13
,, gangrenous,	27
,, mercurial,	18
,, phlegmonous and parenchymatous,	11
,, scorbutic,	19
,, ulcerative,	17
Stomato-mykosis sarcinae,	43
Strawberry tongue,	2
Struma, retropharyngea,	190
Sublingual cysts,	49
,, gland,	57
Submaxillary gland,	57
,, diseases of,	67
Suppurating pharyngitis,	131
Symptomatic ozœna,	242
Syphilis of the mouth,	31
,, of the nose,	247
,, of the pharynx,	172
Syphilitic angina,	172
Taste, nerves of,	4
Testing sense of smell,	274
Therapeutics, general, of nose,	212
,, pharynx,	83
Thrush in children, treatment of,	41, 285
,, of mouth,	36
,, of nose,	253
,, of pharynx,	181
Tissue, diphtheria,	146
Tongue, anomalies of,	7
,, cancer of,	48
,, congenital absence of,	7
,, fur of,	2
,, method of examining,	6
,, muscles of,	5
,, papillæ of,	2
,, paralysis of,	54
,, spasm of,	53
,, structure of,	5
,, superficial excoriation of,	24
,, syphilis of,	32
,, tuberculosis of,	35

	PAGE
Tongue, tumours of,	48
Tongue-tie,	7
Tonsillitis,	96
,, interstitial retracting,	174
Tonsillotomy,	130
,, in pharyngeal phlegmon,	136
Tonsil stones and plugs,	187
Tonsils,	73
,, catarrh of,	96
,, concretions in,	187
,, gumma of,	174
,, hypertrophy of,	116, 122
,, phlegmon of,	133
,, sarcoma of,	188
,, tumours of,	188
Tooth formation in nose,	264
Toxic angina,	95
Trachea, diphtheria of,	152
,, stenosis of, from diphtheria,	159
Tracheal diphtheria,	152
,, wound, diphtheria of,	159
Tracheotomy in diphtheria,	168
,, glossitis,	13
,, goitre,	191
,, pharyngeal phlegmon,	136
Transfusion, Duncan's method,	287
Traumatic ulcers of mouth,	23
Trismus or lockjaw,	53
Tuberculosis of mouth,	35
,, of nose,	250
,, of pharynx,	176
,, of sphenoidal sinuses,	280
,, of tongue,	35
Tumours of mouth,	44
,, of nose,	261
,, of pharynx,	188
,, of tongue,	48
Turbinated bones,	203, 210
,, amputation of ends,	235
,, cavernous spongy tissue covering,	205
,, defects and deformities of,	217
,, in rhinoscopic image,	80
,, occasional presence of a fourth,	203, 217
,, phlegmon of,	245
Tylosis linguæ,	44
Typhoid parotitis,	64
Ulcerative stomatitis,	17
Unna, map-tongue,	25
Urbantschitsch on chorda tympani,	285
Urine in diphtheria,	149
Uvula,	73
,, bifida,	89
,, extravasation of blood into,	184

	PAGE
Uvula, hypertrophy of,	122
,, indications for amputating,	129
Variola of the mouth,	15
,, pharynx,	142
Vasomotor neurosis of pharynx,	195
Vegetations, adenoid,	104
Vertigo in chronic rhinitis,	231
Vesicular stomatitis,	13
Vestibulum oris,	1
Vicarious epistaxis,	256
Voice, effect of chronic pharyngitis on,	121
,, ,, diphtheria on,	149
Von Troeltsch, method of gargling,	286
Water cancer, or noma,	28
Weber's nasal douche,	212
Wharton's duct,	57

	PAGE
Winternitz, experiments on blood pressure,	83
Wolfsrachen, or wolf's jaw,	7, 89
Yeast in oral scurvy,	20
Zaufal's nasal spatula,	211
,, snare for naso-pharyngeal polypi,	270
,, specula,	81, 210
Zenker, thrush fungus in brain,	40
Ziem, treatment of atrophic rhinitis by ventilation,	243
Zittmann's decoction,	34, 285
Zuckerkandl, classification of mucous polypi of antrum,	278
,, on function of spongy tissue,	205

www.ingramcontent.com/pod-product-compliance
Lightning Source LLC
Chambersburg PA
CBHW031903220426
43663CB00006B/750